PROGRESS IN CLINICAL AND BIOLOGICAL RESEARCH

Series Editors
Nathan Back Vincent P. Eijsvoogel Kurt Hirschhorn Sidney Udenfriend
George J. Brewer Robert Grover Seymour S. Kety Jonathan W. Uhr

RECENT TITLES

Vol 210: **Ionic Currents in Development,** Richard Nuccitelli, *Editor*

Vol 211: **Transfusion Medicine: Recent Technological Advances,** Kris Murawski, Frans Peetoom, *Editors*

Vol 212: **Cancer Metastasis: Experimental and Clinical Strategies,** D.R. Welch, B.K. Bhuyan, L.A. Liotta, *Editors*

Vol 213: **Plant Flavonoids in Biology and Medicine: Biochemical, Pharmacological, and Structure–Activity Relationships,** Vivian Cody, Elliott Middleton, Jr., Jeffrey B. Harborne, *Editors*

Vol 214: **Ethnic Differences in Reactions to Drugs and Xenobiotics,** Werner Kalow, H. Werner Goedde, Dharam P. Agarwal, *Editors*

Vol 215: **Megakaryocyte Development and Function,** Richard F. Levine, Neil Williams, Jack Levin, Bruce L. Evatt, *Editors*

Vol 216: **Advances in Cancer Control: Health Care Financing and Research,** Lee E. Mortenson, Paul F. Engstrom, Paul N. Anderson, *Editors*

Vol 217: **Progress in Developmental Biology,** Harold C. Slavkin, *Editor.* Published in two volumes.

Vol 218: **Evolutionary Perspective and the New Genetics,** Henry Gershowitz, Donald L. Rucknagel, Richard E. Tashian, *Editors*

Vol 219: **Recent Advances in Arterial Diseases: Atherosclerosis, Hypertension, and Vasospasm,** Thomas N. Tulenko, Robert H. Cox, *Editors*

Vol 220: **Safety and Health Aspects of Organic Solvents,** Vesa Riihimäki, Ulf Ulfvarson, *Editors*

Vol 221: **Developments in Bladder Cancer,** Louis Denis, Tadao Niijima, George Prout, Jr., Fritz H. Schröder, *Editors*

Vol 222: **Dietary Fat and Cancer,** Clement Ip, Diane F. Birt, Adrianne E. Rogers, Curtis Mettlin, *Editors*

Vol 223: **Cancer Drug Resistance,** Thomas C. Hall, *Editor*

Vol 224: **Transplantation: Approaches to Graft Rejection,** Harold T. Meryman, *Editor*

Vol 225: **Gonadotropin Down-Regulation in Gynecological Practice,** Rune Rolland, Dev R. Chadha, Wim N.P. Willemsen, *Editors*

Vol 226: **Cellular Endocrinology: Hormonal Control of Embryonic and Cellular Differentiation,** Ginette Serrero, Jun Hayashi, *Editors*

Vol 227: **Advances in Chronobiology,** John E. Pauly, Lawrence E. Scheving, *Editors.* Published in two volumes.

Vol 228: **Environmental Toxicity and the Aging Processes,** Scott R. Baker, Marvin Rogul, *Editors*

Vol 229: **Animal Models: Assessing the Scope of Their Use in Biomedical Research,** Junichi Kawamata, Edward C. Melby, Jr., *Editors*

Vol 230: **Cardiac Electrophysiology and Pharmacology of Adenosine and ATP: Basic and Clinical Aspects,** Amir Pelleg, Eric L. Michelson, Leonard S. Dreifus, *Editors*

Vol 231: **Detection of Bacterial Endotoxins With the Limulus Amebocyte Lysate Test,** Stanley W. Watson, Jack Levin, Thomas J. Novitsky, *Editors*

Vol 232: **Enzymology and Molecular Biology of Carbonyl Metabolism: Aldehyde Dehydrogenase, Aldo-Keto Reductase, and Alcohol Dehydrogenase,** Henry Weiner, T. Geoffrey Flynn, *Editors*

Vol 233: **Developmental and Comparative Immunology,** Edwin L. Cooper, Claude Langlet, Jacques Bierne, *Editors*

Vol 234: **The Hepatitis Delta Virus and Its Infection**, Mario Rizzetto, John L. Gerin, Robert H. Purcell, *Editors*

Vol 235: **Preclinical Safety of Biotechnology Products Intended for Human Use**, Charles E. Graham, *Editor*

Vol 236: **First Vienna Shock Forum**, Günther Schlag, Heinz Redl, *Editors*. Published in two volumes: Part A: *Pathophysiological Role of Mediators and Mediator Inhibitors in Shock.* Part B: *Monitoring and Treatment of Shock.*

Vol 237: **The Use of Transrectal Ultrasound in the Diagnosis and Management of Prostate Cancer**, Fred Lee, Richard McLeary, *Editors*

Vol 238: **Avian Immunology**, W.T. Weber, D.L. Ewert, *Editors*

Vol 239: **Current Concepts and Approaches to the Study of Prostate Cancer**, Donald S. Coffey, Nicholas Bruchovsky, William A. Gardner, Jr., Martin I. Resnick, James P. Karr, *Editors*

Vol 240: **Pathophysiological Aspects of Sickle Cell Vaso-Occlusion**, Ronald L. Nagel, *Editor*

Vol 241: **Genetics and Alcoholism**, H. Werner Goedde, Dharam P. Agarwal, *Editors*

Vol 242: **Prostaglandins in Clinical Research**, Helmut Sinzinger, Karsten Schrör, *Editors*

Vol 243: **Prostate Cancer**, Gerald P. Murphy, Saad Khoury, Réne Küss, Christian Chatelain, Louis Denis, *Editors*. Published in two volumes: Part A: *Research, Endocrine Treatment, and Histopathology.* Part B: *Imaging Techniques, Radiotherapy, Chemotherapy, and Management Issues*

Vol 244: **Cellular Immunotherapy of Cancer**, Robert L. Truitt, Robert P. Gale, Mortimer M. Bortin, *Editors*

Vol 245: **Regulation and Contraction of Smooth Muscle**, Marion J. Siegman, Andrew P. Somlyo, Newman L. Stephens, *Editors*

Vol 246: **Oncology and Immunology of Down Syndrome**, Ernest E. McCoy, Charles J. Epstein, *Editors*

Vol 247: **Degenerative Retinal Disorders: Clinical and Laboratory Investigations**, Joe G. Hollyfield, Robert E. Anderson, Matthew M. LaVail, *Editors*

Vol 248: **Advances in Cancer Control: The War on Cancer—15 Years of Progress**, Paul F. Engstrom, Lee E. Mortenson, Paul N. Anderson, *Editors*

Vol 249: **Mechanisms of Signal Transduction by Hormones and Growth Factors**, Myles C. Cabot, Wallace L. McKeehan, *Editors*

Vol 250: **Kawasaki Disease**, Stanford T. Shulman, *Editor*

Vol 251: **Developmental Control of Globin Gene Expression**, George Stamatoyannopoulos, Arthur W. Nienhuis, *Editors*

Vol 252: **Cellular Calcium and Phosphate Transport in Health and Disease**, Felix Bronner, Meinrad Peterlik, *Editors*

Vol 253: **Model Systems in Neurotoxicology: Alternative Approaches to Animal Testing**, Abraham Shahar, Alan M. Goldberg, *Editors*

Vol 254: **Genetics and Epithelial Cell Dysfunction in Cystic Fibrosis**, John R. Riordan, Manuel Buchwald, *Editors*

Vol 255: **Recent Aspects of Diagnosis and Treatment of Lipoprotein Disorders: Impact on Prevention of Atherosclerotic Diseases**, Kurt Widhalm, Herbert K. Naito, *Editors*

Vol 256: **Advances in Pigment Cell Research**, Joseph T. Bagnara, *Editor*

Vol 257: **Electromagnetic Fields and Neurobehavioral Function**, Mary Ellen O'Connor, Richard H. Lovely, *Editors*

Vol 258: **Membrane Biophysics III: Biological Transport**, Mumtaz A. Dinno, William McD. Armstrong, *Editors*

Vol 259: **Nutrition, Growth, and Cancer**, George P. Tryfiates, Kedar N. Prasad, *Editors*

Vol 260: **Management of Advanced Cancer of Prostate and Bladder**, Philip H. Smith, Michele Pavone-Macaluso, *Editors*

Vol 261: **Nicotine Replacement: A Critical Evaluation**, Ovide F. Pomerleau, Cynthia S. Pomerleau, *Editors*

Vol 262: **Hormones, Cell Biology, and Cancer: Perspectives and Potentials**, W. David Hankins and David Puett, *Editors*

Please contact the publisher for information about previous titles in this series.

Hormones, Cell Biology, and Cancer
Perspectives and Potentials

HORMONES, CELL BIOLOGY, AND CANCER
Perspectives and Potentials

Proceedings of a Satellite Symposium of the 1986 American Society for Cell Biology Meeting, Held in Washington, DC, December 7, 1986

Editors

W. David Hankins
Division of Hematology
Armed Forces Radiobiology
Research Institute, and
Laboratory of Chemical Biology
National Institutes of Health
Bethesda, Maryland

David Puett
Department of Biochemistry
The Reproductive Sciences and
Endocrinology Laboratories
University of Miami School of Medicine
Miami, Florida

ALAN R. LISS, INC. • NEW YORK

Address all Inquiries to the Publisher
Alan R. Liss, Inc., 41 East 11th Street, New York, NY 10003

Copyright © 1988 Alan R. Liss, Inc.

Printed in the United States of America

Under the conditions stated below the owner of copyright for this book hereby grants permission to users to make photocopy reproductions of any part or all of its contents for personal or internal organizational use, or for personal or internal use of specific clients. This consent is given on the condition that the copier pay the stated per-copy fee through the Copyright Clearance Center, Incorporated, 27 Congress Street, Salem, MA 01970, as listed in the most current issue of "Permissions to Photocopy" (Publisher's Fee List, distributed by CCC, Inc.), for copying beyond that permitted by sections 107 or 108 of the US Copyright Law. This consent does not extend to other kinds of copying, such as copying for general distribution, for advertising or promotional purposes, for creating new collective works, or for resale.

While the authors, editors, and publisher believe that drug selection and dosage and the specifications and usage of equipment and devices, as set forth in this book, are in accord with current recommendations and practice at the time of publication, they accept no legal responsibility for any errors or omissions, and make no warranty, express or implied, with respect to material contained herein. In view of ongoing research, equipment modifications, changes in governmental regulations and the constant flow of information relating to drug therapy, drug reactions and the use of equipment and devices, the reader is urged to review and evaluate the information provided in the package insert or instructions for each drug, piece of equipment or device for, among other things, any changes in the instructions or indications of dosage or usage and for added warnings and precautions.

LIBRARY OF CONGRESS

Library of Congress Cataloging-in-Publication Data

Hormones, cell biology, and cancer : perspectives and potentials :
   proceedings of a satellite symposium of the 1986 American Society
   for Cell Biology Meeting, held in Washington, D.C., December 7, 1986
   / editors, W. David Hankins, David Puett.
      p.   cm.—(Progress in clinical and biological research ; v.
262)
    Symposium sponsored by the U.S. Cancer Research Council.
    Includes bibliographies and index.
    ISBN 0-8451-5112-6
    1. Cancer—Endocrine aspects—Congresses. 2. Hormone receptors—
Congresses. I. Hankins, W. David. II. Puett, David.
III. American Society for Cell Biology. Meeting (1986 : Washington,
D.C.) IV. U.S. Cancer Research Council. V. Series.
    [DNLM: 1. Cells—physiology—congresses. 2. Hormones—
pharmacodynamics—congresses. 3. Hormones—physiology—congresses.
4. Neoplasms—physiopathology—congresses. W1 PR668E v. 262 / QZ
200 H8125]
RC268.2.H673 1987
616.99'4—dc 19
DNLM/DLC
for Library of Congress   87-34255
                                                                                   CIP

Contents

Contributors	ix
Preface	xiii
Acknowledgments	xv

STEROID HORMONES AND RECEPTORS

From Cell Proliferation to Differentiation: The Steroid Connection
Alan Garen 3

Combination Therapy With Flutamide and [D-Trp^6, DES-GLY-NH$_2^{10}$]LHRH Ethylamide in Stage C and D Prostate Cancer: Today's Therapy of Choice—Rationale and 5-Year Clinical Experience
Fernand Labrie, André Dupont, Lionel Cusan, Alain Bélanger, Michel Giguère, Claude Labrie, Yves Lacourcière, Gérard Monfette, and Jean Emond 11

Aromatase and Aromatase Inhibitors: From Enzymology to Selective Chemotherapy
M.E. Brandt, D. Puett, R. Garola, K. Fendl, D.F. Covey, and S.J. Zimniski . . 65

Estrogens and Antiestrogens Mediate Contrasting Transitions in Estrogen Receptor Conformation Which Determine Chromatin Access: A Review and Synthesis of Recent Observations
Katherine Nelson, John R. van Nagell, Holly Gallion, Elvis S. Donaldson, and Edward J. Pavlik 85

Long-Term Tamoxifen Therapy to Control or to Prevent Breast Cancer: Laboratory Concept to Clinical Trials
V. Craig Jordan 105

Estrogen Receptor in Human Prostate: Assay Conditions for Quantitation
S. Ganesan, N. Bashirelahi, and J.D. Young, Jr. 125

Validation of the Exchange Assay for the Measurement of Androgen Receptors in Human and Dog Prostates
Abdulmaged M. Traish, Donald F. Williams, Neil D. Hoffman, and Herbert H. Wotiz 145

Progestin Effects on Lactate Dehydrogenase and Growth in the Human Breast Cancer Cell Line T47D
Michael R. Moore, Rodney D. Hagley, and Judith R. Hissom 161

The Cellular Response of Human Breast Cancer to Estrogen
G. Wilding, M.E. Lippman, and R.B. Dickson **181**

PEPTIDE HORMONES AND RECEPTORS

Cellular Endocrinology and Cancer
Gordon H. Sato **199**

Epidermal Growth Factor
Barbara Mroczkowski and Graham Carpenter **207**

Use of Analogs of LH-RH and Somatostatin for the Treatment of Hormone Dependent Cancers
Tommie W. Redding and Andrew V. Schally **217**

Multi-Hormonal Regulation of Tyrosinase Expression in B16/C3 Melanoma Cells in Culture
Terry J. Smith and Ellis L. Kline **241**

Hormone Associated Therapy of Leukemia: Reflections
W. David Hankins, Kyung Chin, and George Sigounas **257**

Human B Cell Growth Factor and Neoplasia
Chintaman G. Sahasrabuddhe, Sudhir Sekhsaria, Barbara Martin,
Linda Yoshimura, and R.J. Ford **269**

The Role of the Multichain IL-2 Receptor Complex in the Control of Normal and Malignant T-cell Proliferation
Thomas A. Waldmann and Mitsuru Tsudo **283**

Index **295**

Contributors

N. Bashirelahi, Department of Biochemistry, Dental School and the Department of Surgery, Division of Urology, School of Medicine, University of Maryland, Baltimore, MD 21201 [125]

Alain Bélanger, Departments of Molecular Endocrinology and Medicine, Laval University Medical Center, Quebec G1V 4G2, Canada [11]

M.E. Brandt, The Reproductive Sciences and Endocrinology Laboratories, Department of Biochemistry, University of Miami School of Medicine, Miami, FL 33101 [65]

Graham Carpenter, Department of Biochemistry, Vanderbilt University School of Medicine, Nashville, TN 37232 [207]

Kyung Chin, Division of Hematology, Armed Forces Radiobiology Research Institute, Bethesda, MD 20814 [257]

D.F. Covey, Department of Pharmacology, Washington University School of Medicine, St. Louis, MO 63110 [65]

Lionel Cusan, Departments of Molecular Endocrinology and Medicine, Laval University Medical Center, Quebec G1V 4G2, Canada [11]

R.B. Dickson, Breast Cancer Section, Medicine Branch, National Cancer Institute, National Institutes of Health, Bethesda, MD 20892 [181]

Elvis S. Donaldson, Departments of Obstetrics and Gynecology and Biochemistry, University of Kentucky College of Medicine, Lexington, KY 40536 [85]

André Dupont, Departments of Molecular Endocrinology and Medicine, Laval University Medical Center, Quebec G1V 4G2, Canada [11]

Jean Emond, Departments of Molecular Endocrinology and Medicine, Laval University Medical Center, Quebec G1V 4G2, Canada [11]

K. Fendl, The Reproductive Sciences and Endocrinology Laboratories, Department of Biochemistry, University of Miami School of Medicine, Miami, FL 33101 [65]

R.J. Ford, University of Texas System Cancer Center, M.D. Anderson Hospital and Tumor Institute, Houston, TX 77030 [269]

Holly Gallion, Departments of Obstetrics and Gynecology and Biochemistry, University of Kentucky College of Medicine, Lexington, KY 40536 [85]

S. Ganesan, Department of Biochemistry, Dental School and the Department of Surgery, Division of Urology, School of Medicine, University of Maryland, Baltimore, MD 21201 [125]

Alan Garen, Department of Molecular Biophysics and Biochemistry, Yale University, New Haven, CT 06520 [3]

The numbers in brackets are the opening page numbers of the contributors' articles.

x / Contributors

R. Garola, The Reproductive Sciences and Endocrinology Laboratories, Department of Biochemistry, University of Miami School of Medicine, Miami, FL 33101 **[65]**

Michel Giguère, Departments of Molecular Endocrinology and Medicine, Laval University Medical Center, Quebec G1V 4G2, Canada **[11]**

Rodney D. Hagley, Department of Biochemistry, Marshall University School of Medicine, Huntington, WV 25704 **[161]**

W. David Hankins, Division of Hematology, Armed Forces Radiobiology Research Institute, and Laboratory of Chemical Biology, National Institutes of Health, Bethesda, MD 20205 **[257]**

Judith R. Hissom, Department of Biochemistry, Marshall University School of Medicine, Huntington, WV 25704 **[161]**

Neil D. Hoffman, Boston University School of Medicine, Department of Biochemistry, Boston, MA 02118 **[145]**

V. Craig Jordan, Departments of Human Oncology and Pharmacology, University of Wisconsin Clinical Cancer Center, Madison, WI 53792 **[105]**

Ellis L. Kline, Department of Microbiology, Clemson University, Clemson, SC 29634 **[241]**

Claude Labrie, Departments of Molecular Endocrinology and Medicine, Laval University Medical Center, Quebec G1V 4G2, Canada **[11]**

Fernand Labrie, Departments of Molecular Endocrinology and Medicine, Laval University Medical Center, Quebec G1V 4G2, Canada **[11]**

Yves Lacourcière, Departments of Nuclear Medicine and Urology, Laval University Medical Center, Quebec G1V 4G2, Canada **[11]**

M.E. Lippman, Breast Cancer Section, Medicine Branch, National Cancer Institute, National Institutes of Health, Bethesda, MD 20892 **[181]**

Barbara Martin, University of Texas System Cancer Center, M.D. Anderson Hospital and Tumor Institute, Houston, TX 77030 **[269]**

Gérard Monfette, Departments of Molecular Endocrinology and Medicine, Laval University Medical Center, Quebec G1V 4G2, Canada **[11]**

Michael R. Moore, Department of Biochemistry, Marshall University School of Medicine, Huntington, WV 25704 **[161]**

Barbara Mroczkowski, Department of Biochemistry, Vanderbilt University School of Medicine, Nashville, TN 37232 **[207]**

Katherine Nelson, Departments of Obstetrics and Gynecology and Biochemistry, University of Kentucky College of Medicine, Lexington, KY 40536 **[85]**

Edward J. Pavlik, Departments of Obstetrics and Gynecology and Biochemistry, University of Kentucky College of Medicine, Lexington, KY 40536 **[85]**

David Puett, Department of Biochemistry, The Reproductive Sciences and Endocrinology Laboratories, University of Miami School of Medicine, Miami, FL 33101 **[65]**

Tommie W. Redding, Endocrine, Polypeptide and Cancer Institute, VA Medical Center and Section of Experimental Medicine, Department of Medicine, Tulane University School of Medicine, New Orleans, LA 70146 [217]

Chintaman G. Sahasrabuddhe, University of Texas System Cancer Center, M.D. Anderson Hospital and Tumor Institute, Houston, TX 77030 [269]

Gordon H. Sato, W. Alton Jones Cell Science Center, Inc., Lake Placid, NY 12946 [199]

Andrew V. Schally, Endocrine, Polypeptide and Cancer Institute, VA Medical Center and Section of Experimental Medicine, Department of Medicine, Tulane University School of Medicine, New Orleans, LA 70146 [217]

Sudhir Sekhsaria, University of Texas System Cancer Center, M.D. Anderson Hospital and Tumor Institute, Houston, TX 77030 [269]

George Sigounas, Division of Hematology, Armed Forces Radiobiology Research Institute, Bethesda, MD 20814 [257]

Terry J. Smith, Department of Medicine, State University of New York at Buffalo and the Veterans Administration Medical Center, Buffalo, NY 14215 [241]

Abdulmaged M. Traish, Boston University School of Medicine, Department of Biochemistry, Boston, MA 02118 [145]

Mitsuru Tsudo, The Metabolism Branch, National Cancer Institute, National Institutes of Health, Bethesda, MD 20892 [283]

John R. van Nagell, Departments of Obstetrics and Gynecology and Biochemistry, University of Kentucky College of Medicine, Lexington, KY 40536 [85]

Thomas A. Waldmann, The Metabolism Branch, National Cancer Institute, National Institutes of Health, Bethesda, MD 20892 [283]

G. Wilding, Breast Cancer Section, Medicine Branch, National Cancer Institute, National Institutes of Health, Bethesda, MD 20892 [181]

Donald F. Williams, Boston University School of Medicine, Department of Biochemistry, Boston, MA 02118 [145]

Herbert H. Wotiz, Boston University School of Medicine, Department of Biochemistry, Boston, MA 02118 [145]

Linda Yoshimura, University of Texas System Cancer Center, M.D. Anderson Hospital and Tumor Institute, Houston, TX 77030 [269]

J.D. Young, Jr., Department of Biochemistry, Dental School and the Department of Surgery, Division of Urology, School of Medicine, University of Maryland, Baltimore, MD 21201 [125]

S.J. Zimniski, The Reproductive Sciences and Endocrinology Laboratories, Department of Biochemistry, University of Miami School of Medicine, Miami, FL 33101 [65]

Gordon Sato
Director,
W. Alton Jones Cell
Science Center
Lake Placid, NY

Graham Carpenter
Department of Biochemistry
Vanderbilt University School
of Medicine, Nashville, TN

Tommie Redding
Department of Medicine
Tulane University
School of Medicine
New Orleans, LA

Alan Garen
Department of
Molecular Biophysics and
Biochemistry, Yale University
New Haven, CT

David Puett
Reproductive and Endocrine
Laboratories, University of
Miami School of Medicine
Miami, FL

Fernand Labrie
Head, Department of Molecular
Endocrinology, Laval
University Medical Center
Quebec G1V 4G2, Canada

David Hankins
Armed Forces Radiobiology
Research Institute/
National Institutes
of Health, Bethesda, MD

Preface

This book evolved from a satellite symposium of the 1986 American Society for Cell Biology meeting (Washington D.C.) in which Professor Charles B. Huggins delivered the plenary lecture. His pioneering contributions to the field of hormone-responsive tumors over several decades of laboratory work were instrumental in the development of this area of research. The Symposium was sponsored by the U.S. Cancer Research Council for the express purpose of enhancing our understanding of the hormonal requirements for cell growth and the use of hormone therapy, including combinational therapy, in treating various types of cancer. These topics are treated at the molecular and cellular level; in addition, reports are provided on animal and clinical studies.

The basic premise on which the Symposium and this volume are based is illustrated by the comments of Dr. Gordon Sato, Director of the W. Alton Jones Cell Science Center in Lake Placid, New York, and a speaker at this Symposium. He states that all cells, including transformed cells, require a certain combination of hormones for growth. Elucidating those particular hormones, including growth factors, for each and every cell type, normal and transformed, will aid in clarifying the complex roles governing cell growth and should offer more rational approaches to growth inhibition of certain tumor cells. This approach has proven partially successful in treating certain types of breast and prostatic cancers. We strongly believe that all forms of cancer can respond to selected versions of hormonal or antihormonal therapy, and the individual chapters reinforce this hypothesis.

The book is divided into major sections dealing with: (i) steroid hormones and their receptors, and (ii) peptide hormones and their receptors. There is, however, considerable overlap in the two sections. The various chapters reflect, in some instances, detailed reviews and, in other cases, original research findings.

The editors thank the authors for their contributions and pay special tribute to Professor Huggins and the other Symposium speakers: Drs. Graham Carpenter, Fernand Labrie, Alan Garen, Tommie Redding and Gordon Sato.

Acknowledgments

The editors first wish to thank all the participants in the symposium, "Hormones, Cell Biology, and Cancer," a meeting which suggested the hypothesis that all forms of cancer can be treated by various forms of hormonal or antihormonal treatment.

We were gratified by the response to a call for papers at the symposium and thank all the contributors for their thoughtful, timely and provocative articles.

The symposium was held at the Annual Meeting of the American Society for Cell Biology and sponsored by the United States Cancer Research Council, a non-profit organization formed to promote the concept of hormone-associated cancer therapy. The editors acknowledge both the Society's organizational assistance and the support of the USCRC.

Finally, we would like to pay special tribute to the innovative research in hormone-associated cancer of Professor Charles B. Huggins, whose early findings and continued work have been an ongoing inspiration, both scientifically and spiritually, in exploring the very exciting potential of the hormonal approach to cancer.

STEROID HORMONES AND RECEPTORS

FROM CELL PROLIFERATION TO DIFFERENTIATION: THE STEROID CONNECTION

Alan Garen

Department of Molecular Biophysics and Biochemistry, Yale University, 260 Whitney Avenue, New Haven, Connecticut

Steroid hormones are attracting increasing attention as one of the key components in the complex system controlling cell proliferation during development. Although there is a broad spectrum of diverse physiological effects associated with steroids, there appears to be a general mechanism of steroid action, which selectively regulates the expression of specific target genes. The purpose of this article is to review the evidence from Drosophila indicating that the steroid hormone ecdysone acts both as an essential growth factor for proliferating cells and as a signal for the developmental switch from proliferation to differentiation; also, to suggest an analogous dual role for mammalian steroid hormones in normal development and transformation to malignancy.

Shortly after the blastoderm stage, Drosophila development separates into two major pathways; one is the larval pathway for the differentiation of specialized larval structures, and the other is the imaginal pathway for the formation of the imaginal discs, which subsequently differentiate into specialized adult structures. The two pathways differ markedly during the larval stage: most cells entering the larval pathway develop polytene or polyploid chromosomes and differentiated characteristics, in contrast to cells entering the imaginal pathway which retain diploid chromosomes and relatively undifferentiated characteristics (Fig. 1). The imaginal cells in each disc proliferate at a constant exponential rate during the

DUAL PATHWAYS OF DROSOPHILA DEVELOPMENT

LARVAL PATHWAY
- CELL DIVISION STOPS
- DNA REPLICATION CONTINUES
- CELLS GROW AND DIFFERENTIATE

IMAGINAL PATHWAY
- CELL DIVISION CONTINUES | CELL DIVISION STOPS
- CELLS REMAIN SMALL AND RELATIVELY UNDIFFERENTIATED | CELLS GROW AND DIFFERENTIATE

| EMBRYONIC AND LARVAL STAGES | PUPAL AND ADULT STAGES |

figure 1

larval stage (1). Towards the end of the larval stage most of the imaginal cells, like the larval cells earlier in development, also stop proliferating and enter the pupal stage of differentiation. The signal for the developmental switch from imaginal cell proliferation to differentiation is a sharp increase in titer of the steroid hormone ecdysone, as indicated by two lines of evidence. One involves temperature-shift experiments with the temperature-sensitive mutant ecd1: when ecd1 larvae are shifted from a permissive to a restrictive temperature a few hours before the end of larval stage, there is no increase in ecdysone titer and the imaginal cells continue proliferating for several days, forming greatly enlarged discs which nevertheless are still capable of differentiating when transplanted into wild-type larval hosts (2,3; see fig. 2). The other evidence is that permanent cultures of imaginal cells can be established in vivo by serial transfers of mature imaginal discs into the abdominal cavity of adult female hosts (4), which have about the same ecdysone titer as third-instar larvae before the titer increases (2). When the cultured imaginal cells are transplanted back into a late third-instar larval host, the cells stop proliferating and differentiate synchronously with the imaginal discs of the host during its pupal stage (4). Differentiation of cultured imaginal cells can also be elicited in an adult

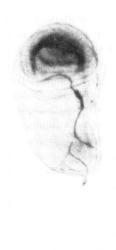
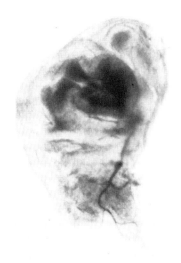

figure 2
Effect of blocking an ecdysone signal on the proliferation of <u>Drosophila</u> imaginal cells.

The left panel shows an imaginal wing disc from a fully grown wild-type larva; the disc contains about 50,000 imaginal cells, which is the maximum population size reached during normal development. The right panel shows an enlarged imaginal wing disc from an <u>ecd^1</u> mutant larva, which was shifted to a high temperature (restrictive) growth condition shortly before pupation and kept four days. This treatment of <u>ecd^1</u> larvae blocks the ecdysone signal which is needed to switch development from the larval to the pupal stage, and the imaginal cells continue proliferating within the enlarged discs.

female host by injecting ecdysone (5). Thus imaginal cells can either be maintained indefinitely as a proliferating population when the ecdysone titer is low, or can be induced to switch from proliferation to differentiation by increasing the ecdysone titer. The developmental switch in the larval pathway, which occurs early in embryogenesis, also appears to be induced by an ecdysone signal, since there is a sharp increase in ecdysone titer at about the same time (6). In *in vitro* cultures of *Drosophila* cells (whose lineage is uncertain), the cells respond rapidly and irreversibly when the culture medium is supplemented with physiological concentrations of ecdysone: cell proliferation stops and is usually followed by morphological changes (7).

The ecdysone titer is continuously adjusted, both upwards and downwards, throughout *Drosophila* development (6), and these adjustments appear to be essential for the normal progress of development and for the fertility of the adult fly (2). There is not much known about the mechanism of ecdysone regulation. The imaginal cells are not involved, since larval development proceeds normally, including the appearance of the ecdysone signal towards the end of the larval stage, in autonomous *discless* mutants which lack all imaginal discs (8-10). The principal role of ecdysone, and probably its sole role, is as a hormonal cofactor in the selective regulation of gene expression, causing either induction or repression of transcription (11-16). Accordingly, cell proliferation could depend on ecdysone-induced activation of genes required for the mitotic cycle, and the switch from proliferation to differentiation on ecdysone-induced repression of some or all of those genes.

Since evolution tends to conserve basic mechanisms of development, as dramatically illustrated by the homeotic genes (17), it seems reasonable to consider that other organisms might also use steroid hormones to induce a switch from cell proliferation to differentiation at the appropriate stages of development. Although in *Drosophila* most cells, and perhaps all, respond to the same steroid, ecdysone, higher organisms contain numerous steroids which show cell-type specificity.

The control of cell proliferation is a central problem not only of normal development but also of malignant development. There is convincing evidence that mutations in certain genes are a necessary step in the progression towards cancer (18; 19). Since mutations generally occur in dividing cells, the developmental switch that stops cell proliferation is particularly relevant to an understanding of the origins of cancer.

During the proliferative phase of development, when the cell populations that will differentiate into the various parts of an organism are being generated, somatic mutations could occur that impair or block the cellular response to a hormonal signal that stops proliferation. Such mutant cells probably would be benign initially, but would generate increasingly large clones of dividing cells within the differentiated population. Those cells would be susceptible to additional mutations, including tumorigenic mutations, which could eventually result in a transformation to malignancy. However, as long as the cells retain the capacity to undergo the switch from proliferation to differentiation, perhaps when exposed to a stronger hormonal signal, conditions might be found for inducing the switch and consequently preventing or reversing the onset of the tumor.

The hormonal signals that regulate cell division include, besides steroids, peptide growth factors such as EGF, TGF and PDGF (20, 21). Evidence is accumulating that some oncogenes are altered or amplified growth factor structural genes, or altered growth factor receptor genes, which cause abnormal stimulation of cell division (20, 21). The recent finding that the oncogene v-erb-A, and its normal cellular homolog c-erb-A, are structurally related to genes encoding receptors for steroid hormones (22) provides support for a central role of steroid hormones in the process of malignant transformation.

One of the clinical approaches to the treatment of human breast cancer involves the steroid hormone estrogen, which appears to be an essential growth factor for tumors containing a high titer of the estrogen receptor (23). Two strategies which have opposite aims are currently used: one aim is to deprive the tumor cells of estrogen, either by surgically blocking its biosynthesis or by administering estrogen analogs to block access to the receptors; the other aim is to raise the estrogen titer by administering large doses of the hormone (24). The rationale for the first treatment is clear, given the evidence that estrogen is needed to support the growth of most breast tumors. Although there has not been a clear rationale offered for the second treatment, it fits well with the proposed role of steroids as hormonal signals for the switch from cell proliferation to differentiation. The tumors affected by the second treatment could have a defective mechanism for responding to the hormonal signal, requiring higher than normal titers of estrogen to stop proliferation. Both treatments appear to elicit "objective tumor regression" in about 60% of the tumors which are classified as estrogen receptor positive (24).

Another relevant finding concerns the beneficial effects of using contraceptive pills, which contain a combination of estrogen and progestin, on the incidence of certain cancers of the female reproductive organs, notably ovarian and endometrial cancers (25, 26). The incidence of ovarian cancer among women who had used the pill, as compared to those who had not, was reduced to about half, even when more than 10 years had elapsed since the date of the last use. One interpretation of this remarkable result is that raising the titer of the steroids induces a permanent switch from proliferation to differentiation in susceptible ovarian cells.

Although we are still far from understanding the control of cell proliferation during normal development, and the loss of control which leads to cancer, a sharper focus on the role of steroid hormones seems justified by the limited information on hand. Further progress depends on the identification and characterization of the cellular components which respond to steroid signals, particularly the signal initiating an irreversible developmental switch from cell proliferation to differentiation. The problem is difficult and complex, involving steroid receptors and their in-target genes and encoded proteins (22). There are strong incentives to proceed, not only to fill a major gap in our understanding of basic developmental processes, but also because the information obtained could lead to a more effective hormonal approach to cancer therapy and prevention. Most of the efforts have been directed towards blocking the growth-stimulating action of steroids, with notable results for prostate and breast cancers. The complementary approach of administering the appropriate steroid to induce a switch from proliferation to differentiation deserves similar attention, and offers hope of applying normal developmental signals to control abnormal development.

REFERENCES

1. Martin, PF "Direct Determination of the Growth Rate of Drosophila Imaginal Discs" J. Expt. 2001. 222, 97 (1982).
2. Garen, A, Kauvar, L and Lepesant, JA "Roles of Ecdysone in Drosophila Development" Proc. Natl. Acad. Sci. USA 74, 5099 (1977).
3. Garen, A and Lepasant, JA "Hormonal Control of Gene Expression and Development by Ecdysone in Drosophila" in Gene Regulation by Steroid Hormones (AK Roy and JH Clark, editors): 255-262 (1980).
4. Hadorn, E. "Problems of Determination and Transdetermination" in Genetic Control of Differentiatio, Brookhaven Symp. Biol. 18: 148 (1965).
5. Postlethwait, JH and Schneiderman, HA "Induction of Metamorphosis by Ecdysone Analogues: Drosophila Imaginal Disc Cultured in vivo" Biol. Bull. 138: 47 (1970).
6. Cherbas, P, Cherbas, L, Demetri, G, Mantenffelcymborowska, C, Savakis, C, Yonger, CD and Williams, CM "Ecdysteroid Hormone Effects on a Drosophila Cell Line" in Gene Regulation by Steroid Hormones (AR Roy and JH Clark, editors): 278-308 (1980).
7. Richards, G "The Radioimmune Assay of Ecdyeteroid Titers in "Drosophila melanogaster" Mol. Cell Endocrinology 21: 181 (1981).
8. Shearn, A, Rice, T, Garen, A and Gehring, W. "Imaginal Disc Abnormalities in Lethal Mutants of Drosophila" Proc. Natl. Acad. Sci. USA 68: 2594 (1971).
9. Shearn, A and Garen, A "Genetic Control of Imaginal Disc Development in Drosophila" Proc. Natl. Acad. Sci. USA 71: 1393 (1974).
10. Stewart M, Murphy, C and Fristrom, JW "The Recovery and Preliminary Characterization of X Chromosome Mutants Affecting Imaginal Discs of Drosophila melanogaster" Develop. Biol. 27: 71 (1972).
11. Ashburner, M "Sequential Gene Activation by Ecdysone in Polytene Chromosomes of Drosophila melanogaster I. Dependence upon Ecdysone Concentration" Develop Biol. 34: 47 (1973).
12. Gronmeyer, H and Pongs O "Localization of Ecdysterone on Polytene Chromosomes of Drosophila melanogaster" Proc. Natl. Acad. Sci. USA 77: 2108 (1980).
13. Lepesant, JA, Kejzlarova-Lepesant, J and Garen, A "Ecdysone-inducible Functions of Larval Fat Bodies of Drosophila" Proc. Natl. Acad. Sci. USA 75: 5570 (1978).

14. Meyerowitz, E and Hogness, DS "Molecular Organization of a Drosophila Puff Site that Responds to Ecdysone" Cell 28: 165 (1980).
15. Lepesant, JA, Levine, M, Garen, A, Lepesant-Kejzlarova, J, Rat, L and Somme-Martin, G "Developmentally Regulated Expression in Drosophila Larval Fat Bodies" J. Mol. Applied Genetics 1: 371, (1982)
16. Nakanishi, Y and Garen A "Selective Gene Expression Induced by Ecdysterone in Cultured Fat Bodies of Drosophila" Proc. Natl. Acad. Sci. USA 80: 2971 (1983).
17. Manley, JL and Levine, MS "The Homeo Box and Mammalian Development" Cell 43: 1 (1985).
18. Klein, G and Klein, E "Evolution of Tumors and the Impact of Molecular Oncology" Nature 315: 190 (1985).
19. Gateff, E "Malignant Neoplasms of Genetic Origin in Drosophila" Science 200: 1448 (1978).
20. Sporn, MB and Roberts, AB "Autocrine Growth Factors and Cancer" Nature 313: 745 (1985).
21. Heldin, CH and Westermark, B "Growth Factors: Mechanism of Action and Relation to Oncogenes" Cell 37: 9 (1985).
22. Bishop, JM "Oncogenes as Hormone Receptors" Nature 321: 112 (1986).
23. McGuire, WL, Carbone, PP, Sears, ME, and Escher, GC "Estrogen Receptors in Human Breast Cancer: an Overview" in Estrogen Receptors in Human Breast Cancer (McGuire and Carbone, eds. New York, Raven Press) 17 (1975).
24. Legha, SS, Davis HL and Muggia FM "Hormonal Therapy of Breast Cancer: New Approaches and Concepts" Ann. Internal Medicine 88: 69 (1978).
25. Ory, HW "Oral Contraceptive Use and the Risk of Ovarian Cancer" J. Amer. Med. Assoc. 249: 1596 (1983).
26. Ory, HW "Oral Contraceptive Use and the Risk of Endometrial Cancer" J. Amer. Med. Assoc. 249: 1600 (1983).

COMBINATION THERAPY WITH FLUTAMIDE AND [D-Trp^6, DES-GLY-NH$_2^{10}$]LHRH ETHYLAMIDE IN STAGE C AND D PROSTATE CANCER: TODAY'S THERAPY OF CHOICE - RATIONALE AND 5-YEAR CLINICAL EXPERIENCE

Fernand Labrie, André Dupont, Lionel Cusan, Alain Bélanger, Michel Giguère, Claude Labrie, Yves Lacourcière, Gérard Monfette and Jean Emond

Departments of Molecular Endocrinology, Medicine, Nuclear Medicine and Urology, Laval University Medical Center, Quebec G1V 4G2, Canada

INTRODUCTION

Following the discovery by Huggins and his colleagues in 1941 that testicular androgens exert a stimulatory effect on prostate cancer growth, the most significant observation is likely to be the one indicating that the adrenals also secrete important amounts of androgens and the marked heterogeneity of androgen sensitivity of cancer. In fact, human adrenals secrete large amounts of precursor steroids which are converted into active androgens in peripheral tissues (including the prostate), thus providing 40 to 50% of total androgens in adult men. The action of these androgens remaining after castration can be inhibited in the prostatic cancer tissue by administration of a pure antiandrogen which, in addition, decreases the local concentration of dihydrotestosterone (DHT). Recent data have also shown that the castration levels of serum testosterone left in men after castration have an important stimulatory activity on the growth of androgen-sensitive normal as well as cancer tissues. In addition, cancer cells have markedly different requirements for androgens, some cell clones being able to grow in the presence of minimal amounts of androgens, a situation which requires more complete androgen blockade and more potent antiandrogens for inhibition of growth. Among the compounds recommended as antiandrogens, a most unexpected finding is that many of those drugs are devoid of any antiandrogenic activity. In fact, medroxyprogesterone acetate, chlormadinone acetate and megestrol acetate have androgenic activity but do not exert any inhibitory effect on the peripheral action of DHT in prostatic tissue. These compounds should not be classified as antiandrogens. Cyproterone acetate, on the other hand, is a mixed agonist-antagonist. The only compounds showing pure antiandrogenic activity are Flutamide and its analogues.

The above-mentioned fundamental and clinical data, especially those related to the contribution of the adrenals as stimulators of prostate cancer, stress the need for a more complete blockade of androgens of both testicular and adrenal origins in order to exert a maximal inhibitory effect on cancer growth. In agreement with this strategy, we have performed clinical studies in previously untreated stage D2 and C prostate cancer patients with the combination therapy using the LHRH agonist [D-Trp^6, des-Gly-NH$_2$^{10}]LHRH ethylamide and the antiandrogen Flutamide.

One hundred ninety-nine patients with clinical stage D2 prostate cancer who had not received previous endocrine therapy or chemotherapy were treated with the combination therapy using the pure antiandrogen Flutamide and the LHRH agonist [D-Trp^6]LHRH ethylamide for an average of 26 months (3 to 59). The objective response to the treatment was according to the criteria of the US NPCP. There was a 5.7-fold increase (26.3 versus 4.6%) in the percentage of patients who achieved a complete response as compared to the results obtained in 5 recent studies limited to removal (orchiectomy) or blockade (DES or Leuprolide) of testicular androgens. Only 12 of the 186 evaluable patients (6.5%) did not show an objective positive response at the start of the combination therapy as compared to an average of 18% in the same 5 studies using monotherapy. The duration of response was also significantly improved in the patients who received the combination therapy while the death rate was decreased by approximately 2-fold during the first 4 years of treatment. In fact, while an approximately 50% death rate is observed at 2 years in all studies using monotherapy, the same 50% death rate is delayed by 2 years in the present study. It should be mentioned that at the time of relapse under combination therapy, the treatment is continued and, in addition, further blockade of adrenal androgen secretion is achieved with aminoglutethimide. The marked (5.7-fold) improvement in the rate of complete objective responses coupled with the 3-fold decrease in the number of non responders, the increased duration of the positive responses and the 2-fold decrease in the death rate during the first 4 years of treatment are obtained with the combination therapy using Flutamide and castration, thus improving the quality and duration of life with no or minimal side effects. By blocking the androgen receptors in the prostatic cancer tissue, the antiandrogen decreases the action of the androgens of adrenal origin and thus inhibits the growth of a large number of tumors which, otherwise, would continue to be stimulated by the adrenal androgens left after medical or surgical castration.

Since the dual combination therapy is so highly successful in the control of local growth of prostate cancer, we felt important to start therapy at an earlier stage of the disease. For this purpose, sixty- seven previously untreated patients presenting with clinical stage C prostatic carcinoma with no evidence of distant metastases received the same combination therapy for an average duration of treatment of 23.5 months. Only 5 patients have so far shown treatment failure with a probability of continuing response of 91.8% at 2 years. Local control was achieved rapidly in all except one patient. Urinary obstruction and hydronephrosis were corrected in all cases. When comparing to recent data obtained after single endocrine therapy (orchiectomy or estrogens) or radiotherapy, the rate of treatment failure at 2 years is 3.5-fold lower after combination therapy (8.2%) than monotherapy (28.4%). The probability of death at 2 years following start of the combination therapy is 6.5% while it is on average at 22.2% (3.4-fold higher) in the other available studies. By blocking the androgen receptors in the prostatic cancer tissue, the antiandrogen decreases the action of the androgens of adrenal origin and thus inhibits the growth of most of the tumors which, otherwise, remain stimulated by the adrenal androgens left after medical or surgical castration. These advantages are obtained with no or minimal side effects, thus preserving a good quality of life. The high efficacy of the combination therapy in controlling local disease indicates the advantages of early treatment.

A major problem facing the treatment of prostate cancer is the choice of treatment for all those who did not respond to standard therapy and those who relapse after a temporary response to orchiectomy, estrogens or LHRH agonists alone.

Following evidence that adrenal androgens play a role in the progression of cancer in at least a large proportion of these patients, we have administered the combination therapy with Flutamide to a large group of relapsing patients. Two hundred nine patients with biopsy-proven stage D2 prostatic adenocarcinoma showing disease progression after orchiectomy or treatment with DES or an LHRH agonist alone received the combination therapy with the pure antiandrogen Flutamide. In patients treated with DES, the estrogen was replaced by the LHRH agonist [D-Trp^6]LHRH ethylamide. Objective response to therapy was assessed according to the criteria of the US NPCP. Thirteen patients (6.2%) had a complete response to treatment while partial and stable responses were achieved in 20 (9.6%) and 39 (18.7%) patients, respectively, for a total objective response rate of 34.5%. The mean duration of response was 24 months. While, in the non responders, the median survival was 8.1 months with a 17% probability of survival at 2 years, the probabilities of survival at 2 years of the patients who showed partial and stable responses were 87 and 67%, respectively. All patients who achieved a complete response are still alive. Considering its excellent tolerance, the combination therapy with Flutamide and castration (surgical or LHRH agonist) appears to be the treatment of choice for prostate cancer patients in relapse after standard endocrine therapy.

A. INTRODUCTION

Prostate cancer is the second leading cause of death due to cancer in men and the first cause in men aged 60 years or more (Silverberg and Lubera, 1983). Due to the presence of bone metastases in the majority of patients at the time of diagnosis, the possibility of treatment of the primary tumor by surgery and/or radiotherapy is limited to a small proportion of cases, while, for all others, hormonal therapy is the only alternative (Labrie, 1984; Williams and Blooms, 1984; Murphy et al., 1983). Although potentially useful in the future with the availability of other drugs, chemotherapy has been disappointing (Murphy et al., 1983).

The most promising advance in the treatment of prostatic cancer has clearly been the demonstration of the role of testicular androgens by Huggins and his colleagues in 1941 (Huggins and Hodges, 1941). These observations opened a new era in the treatment of prostate cancer and were based on the following straightforward rationale: "In many instances, a malignant prostatic tumor is an overgrowth of adult epithelial cells. All known types of adult prostatic epithelium undergo atrophy when androgenic hormones are greatly reduced in amount or inactivated. Therefore, significant improvement should occur in the clinical condition of patients with far advanced prostate cancer subjected to castration or estrogen administration" (Huggins, 1947). Since these observations of Huggins and his colleagues (Huggins and Hodges, 1941; Huggins, 1947), orchiectomy and treatment with estrogens have been the cornerstone of the management of advanced prostate cancer (Paulson, 1978).

These two approaches cause improvement for a limited time interval in 60 to 80% of cases, thus leaving 20 to 40% of the patients without improvement of their disease (Murphy et al., 1983; Nesbit and Baum, 1950; Jordan et al., 1977; Mettlin et al., 1982). Moreover, progression of the cancer usually occurs within 6 to 24 months in all those who initially responded (Resnick and Grayhack, 1975) with a median life expectancy of only six months (Johnson et al., 1977; Slack et al., 1984). In addition to the uncertain improvement in survival, orchiectomy is often psychologically unacceptable while estrogens cause serious side effects such as gynecomastia, fluid retention, myocardial ischaemia and thromboembolism. By themselves, these side effects of estrogens have been reported to cause death in 15% of patients during the first year of treatment (Glashan and Robinson, 1981). The side effects of the two standard forms of hormonal therapy and their questionable influence on survival left most physicians undecided about the real benefits of hormonal therapy. There was thus the clear need for more efficient and better tolerated therapies.

The finding that agonists of luteinizing hormone-releasing hormone (LHRH) cause a blockade of testosterone secretion in experimental animals (Labrie et al., 1978; Labrie et al., 1980) accompanied by a loss in prostate weight offered the possibility of replacing orchiectomy and estrogens for the treatment of prostate cancer. In fact, in men, following a transient period of stimulation, serum testicular androgens are reduced to castration levels during chronic treatment with the well tolerated LHRH agonists (Labrie et al., 1980; Faure et al., 1982; Warner et al., 1973; Waxman et al., 1983). However, although LHRH agonists make castration more acceptable than orchiectomy and, especially, free of the serious side effects of estrogens, one cannot expect to improve the prognosis of prostate cancer beyond the results previously achieved with orchiectomy since their effect is also limited to the blockade of testicular androgens (Labrie et al., 1980; Labrie et al., 1982; Labrie et al., 1985a).

Before describing the clinical results obtained in advanced prostate cancer using the combination therapy, it seems important to review some pertinent data which demonstrate the major importance of adrenal androgens in prostate cancer in men as well as some biological properties of androgen-sensitive normal and cancer tissues. A summary of the characteristics of the commercially available antiandrogens will also be presented. Then, a proposal for the most appropriate anti hormonal therapy will be presented and our 5-year clinical experience with stage C and D prostate cancer will be summarized.

B- HIGH CONCENTRATIONS OF DHT REMAIN IN THE PROSTATE CANCER TISSUE FOLLOWING CASTRATION

As illustrated in Fig. 1, a high level of the active androgen dihydrotestosterone (DHT) remains in the prostatic cancer tissue following castration. Although orchiectomy, estrogens, or LHRH agonists (through blockade of bioactive LH) cause a 90 to 95% reduction in serum testosterone (T) levels (Labrie et al., 1980; Faure et al., 1982; Warner et al., 1973; Waxman et al., 1983; Labrie et al., 1982; Labrie et al., 1985), a much smaller effect is observed on the really meaningful parameter of androgenic action, namely the concentration of DHT in the prostatic cancer tissue. Of major importance, however, is the finding that the addition of Flutamide to medical castration achieved with the LHRH agonist, [D-Trp^6, des-Gly-NH$_2$^{10}]LHRH ethylamide, causes a reduction of the intraprostatic concentration of DHT to a value below detection limits (0.2 ng/g tissue). Such data indicate the efficacy of flutamide in displacing DHT from the androgen receptor, thus preventing the action of DHT on the androgen receptor and making the steroid readily available for transformation into inactive metabolites and excretion.

Measurements of T and DHT levels in the serum have little or no value except as an index of testicular activity. In fact, the intraprostatic DHT concentration is the only significant parameter which illustrates the level of the active androgen at its site of action in the prostatic cancer tissue itself. Unexpectedly, high concentrations of DHT and 5α-androstane-3α,17β-diol, 3α-DHT) have been found in prostatic carcinoma after orchiectomy or DES treatment (Farnsworth and Brown, 1976; Geller et al., 1984; Bélanger et al., 1986). This is best illustrated in Figure 1 where it can be seen that although orchiectomy causes a 90% to 95% fall in serum T level, the intraprostatic concentration of the potent androgen DHT is reduced by only about 60%.

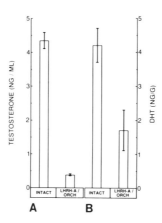

Fig. 1. Effect of castration on the serum levels of T (A) and on the concentration of the active androgen DHT remaining in prostatic cancer tissue after castration (B). Note the relatively small effect (approximately 50%) of castration on intraprostatic DHT concentration as compared to the 90% fall in serum T.

Based on the intraprostatic levels of DHT measured after castration, estrogen treatment or treatment with LHRH agonists alone, it can be concluded that the testes and adrenals are of approximately equal importance in providing androgens which stimulate prostate cancer growth. Such data stress the absolute need to neutralize the action of the androgens of both testicular and adrenal origins for an efficient treatment of prostate cancer. Such high levels of

DHT remaining after surgical or medical castration are more than likely to have an important stimulatory effect on cancer growth. Clearly, with the knowledge of these data, a treatment limited to the blockade of androgens of testicular origin only partially relieves the cancer from androgenic stimulation. Moreover, measurements of serum levels of T and DHT, although being excellent parameters of testicular steroidogenic activity, cannot be used as an index of androgenic activity in the prostatic cancer tissue.

C- CLINICAL EVIDENCE THAT ANDROGEN-HYPERSENSITIVE TUMORS REMAIN AFTER CASTRATION AND DURING RELAPSE

That androgen-sensitive cancer cells remain active after surgical castration or high doses of estrogens is clearly illustrated by the finding that 33% to 39% of patients already castrated or treated with estrogens showed a positive response to the pure antiandrogen Flutamide (Sogani et al., 1975; Stoliar and Albert, 1974). Moreover, after adrenal androgen suppression with aminoglutethimide in patients who had become refractory to orchiectomy and exogenous estrogens, a favorable response was observed in three of seven patients (Sanford et al., 1977). In a similar study, Robinson and coworkers found palliation in 50% of patients (Robinson et al., 1974).

The first bilateral adrenalectomy in prostatic cancer was performed by Huggins and Scott (1945) with appreciable success despite the lack of substitution therapy. Subsequently, bilateral adrenalectomy and hypophysectomy were used in advanced prostatic cancer with a significant rate of remission in previously castrated patients or those already treated with estrogens. In fact, bilateral adrenalectomy has been found to be associated with palliation in 20% to 70% of patients with advanced prostatic carcinoma who had become refractory to castration or estrogen therapy (Sanford et al., 1977; Labrie et al., 1985a). Surgical hypophysectomy has also been found to improve transiently the disease in about 50% of patients (for review, see Labrie et al., 1985a). The benefits of additional androgen blockade in 19/08/87g patients can only be explained by the action of androgens of adrenal origin which remain at a significant level in the prostatic cancer tissue after castration.

The common belief that patients in relapse after castration or treatment with estrogens have exclusively "androgen-insensitive" tumors should be abandonned. In fact, it is mostlikely that androgen-sensitive tumors are present at all stages of prostate cancer in all patients and that optimal androgen blockade should be performed in all cases. Instead of being "androgen-insensitive", most of (if not all) the tumors which continue to grow after castration are androgen-hypersensitive. These tumors are able to grow in the presence of the androgens of adrenal origin left after castration. Inhibition of the growth of these tumors requires further androgen blockade. This affirmation is supported by convincing clinical data as well as by well-established fundamental observations (Labrie et al., 1985a; Labrie and Veilleux, 1986).

Why aim for full androgen blockade? As mentioned above, this is necessary because of the presence of androgen- sensitive tumors in almost all (if not all) patients, even at the time of relapse after castration. The observation by Fowler and Whitmore (1981) of a rapid and severe exacerbation, in 33 of 34 patients in relapse, within the first days of testosterone administration, is an extremely convincing demonstration that prostate cancer remains almost (if not) always androgen-sensitive. Moreover, as mentioned above, a second objective response is well known to be obtained in 30-60% of patients upon addition of further androgen blockade (antiandrogen, adrenalectomy, hypophysectomy, aminoglutethimide) after failure to respond to a first hormonal therapy or relapse (for review, see Labrie et al., 1985a).

D- WIDE RANGE OF SENSITIVITIES OF TUMOR CELLS TO ANDROGENS

Clinical evidence in prostate cancer clearly indicates a marked heterogeneity of the sensitivity to androgens and the development of resistance to standard hormonal therapy. These data pertain to the failure of standard hormonal therapy in an important proportion (20 to 40%) of previously untreated patients while a positive response can be obtained in 90 to 95% of patients by further blockade of androgens (for review, see Labrie et al., 1985a). In fact, while 20-40% of patients do not show a response to standard endocrine therapy (orchiectomy or estrogens), the response obtained in an important proportion of them (20 to 60%) by the addition of adrenalectomy, hypophysectomy or flutamide clearly demonstrates that tumors were left growing under standard hormonal therapy which can be blocked by further androgen blockade. There is thus heterogeneity of androgen sensitivity among tumors present in the same patient as well as between different patients.

Knowledge of the heterogeneity of androgen-sensitivity in prostate cancer in men stresses the importance of studying in detail the phenomenon of heterogeneity and development of "treatment resistance" in originally androgen-sensitive cells. An excellent model for such studies is the androgen- sensitive Shionogi mouse mammary carcinoma 115 (SC-115), a tumor showing rapid growth both in vivo and in tissue culture. We have used this model to investigate the heterogeneity of androgen sensitivity in different clones obtained from a tumor as well as the changes of androgen sensitivity which occur during long-term culture of cloned cells. The data obtained clearly show that a wide range of androgen sensitivities are found in the clones derived from the original tumor and that, furthermore, heterogeneity rapidly develops during tissue culture of cloned cells (Fig. 2, Labrie and Veilleux, 1986). The data are analogous to the situation in human prostate cancer where a population of hypersensitive cells can continue to grow in the presence of the androgens of adrenal origin left after castration (medical or surgical). As illustrated in Fig. 2, the 3 clones obtained from a Shionogi mouse mammary tumor show marked heterogeneity of sensitivity to DHT action. While the original tumor shows a EC_{50} value of DHT action at 0.9 nM, the EC_{50} values of DHT action in clones A, B and C were calculated at 0.024, 0.15 and 30 nM, respectively. There is thus a 1250-fold difference in sensitivity to DHT between clones A and C obtained from a single tumor (Labrie and Veilleux, 1986).

Although the origin of tumors is believed to be monoclonal (Dexter and Calabresi, 1982), it is clear that most, if not all, advanced tumors are composed of mixed populations of cells having a wide range of phenotypes. That heterogeneity of androgen sensitivity analogous to the one described in this report exists in human prostate cancer is unequivocally demonstrated by the clinical data showing a 30 to 60% response to adrenalectomy, hypophysectomy, Flutamide or aminoglutethimide in patients who relapse after orchiectomy or treatment with estrogens (Labrie et al.,

1985a). Such a response to further androgen blockade in patients already castrated can only be explained by the presence in these patients of prostatic tumors which were still growing in the androgenic environment provided by the adrenal androgens remaining after medical or surgical castration.

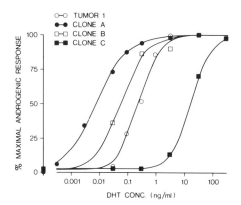

Fig. 2. Effect of increasing concentrations of DHT on the androgenic response (DNA content) in 3 clones obtained from a Shionogi mouse mammary tumor. In order to facilitate visualization of differences in androgen sensitivity, all data are expressed as a percentage of the maximal response to DHT (Labrie and Veilleux, 1986). There were also marked differences in basal growth in the absence of DHT as well as in the maximal response to DHT between the different clones (Labrie and Veilleux, 1986).

The situation in human prostate cancer thus appears to be analogous to the data obtained in the model Shionogi tumor where a proportion of clones show supersensitivity to DHT with half-maximal growth at a concentration of DHT as low as 0.008 ng/ml (0.024 nM) (Fig. 2). Since the castration levels of serum DHT range between 0.04 and 0.08 ng/ml (Labrie et al., 1985a), it is clear that such hypersensitive tumors can continue to grow at a maximal rate following castration. For these androgen-hypersensitive tumors, castration levels of DHT are sufficient to maintain a maximal growth rate.

E- PROPERTIES OF STEROIDAL AND NON-STEROIDAL ANTIANDROGENS

The well recognized sensitivity of prostate cancer to androgens has stimulated the development of compounds, called antiandrogens, which can antagonize or prevent the action of androgens in target tissues (Hamada et al., 1963; Neri et al., 1967). The interest for these compounds is strengthened by the recent observation that the blockade of androgens of both testicular and adrenal origins by combining a pure antiandrogen with castration at the start of treatment of advanced prostate cancer yields a higher response rate, a more prolonged disease-free period and an improved survival while maintaining a good quality of life (Labrie et al., 1982; Labrie et al., 1985a; Labrie et al., 1986). This combination therapy has become necessary following the recognition that adrenal steroids contribute approximately 50% of all androgens present in the prostatic cancer tissue (for review, see Labrie et al., 1985a). Antiandrogens offer an alternative to hypophysectomy or adrenalectomy without the complications of glucocorticoid replacement therapy.

Since the unique aim of endocrine therapy in prostate cancer is to block androgens, the ideal antiandrogen should be a compound having potent antiandrogenic activity while being devoid of any androgenic, glucocorticoid, progestational, estrogenic or any other hormonal or antihormonal action (Dorfmann, 1971). It thus becomes of major importance to assess in detail the properties of available drugs in order to obtain the fundamental data required for making the best choice of antiandrogen. The interest of this study is strengthened by the recent finding that megestrol acetate, a progestin derivative currently used for the treatment of prostatic cancer (Johnson et al., 1981) is a weak androgen completely devoid of any antiandrogenic activity (Poyet and Labrie, 1985) while cyproterone acetate shows stimulatory androgenic effects in many systems (Neri et al., 1967; Poyet and Labrie, 1985; Mowszowicz et al., 1974; Graf et al., 1974; Tisell and Salander, 1975), including stimulation of the growth of androgen-sensitive tumors (Noguchi et al., 1985).

The effect of treatment with 200 µg/day of medroxyprogesterone acetate, cyproterone acetate, chlormadinone acetate or megestrol acetate on size (cm^2/mouse) of the androgen-sensitive Shionogi SC-115 tumor is illustrated in Figure 3. It can be seen that medroxyprogesterone acetate was the most potent stimulator of tumor size followed by cyproterone acetate while megestrol acetate and chlormadinone acetate were much less active and Flutamide was the only compound which did not show a significant stimulatory effect on growth of this androgen-sensitive tumor. In Flutamide-treated animals, although many animals developed palpable tumors, the average tumor size was 0.69 ± 0.23 cm^2/mouse, a value not significantly different from the control group measured at 0.52 ± 0.16 cm^2 on day 25 of the experiment. The androgenic activity of the synthetic progestins is also well illustrated in Figs 4 and 5, where all the compounds tested, except Flutamide, stimulated ventral prostate growth and prostatic ornithine decarboxylase (ODC) activity in castrated rats.

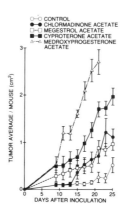

Fig. 3. Effect of treatment with the indicated synthetic compounds (250 µg, twice daily) on size (cm^2/mouse) of the androgen-sensitive Shionogi SC-115 tumor in intact female mice. Treatment was started on the day of the inoculation of the tumor (day 0). Control animals were injected with the vehicle only. Results are presented as means ± SEM.

Antiandrogens must be strictly defined as compounds which inhibit androgen action at the target tissue level, thus excluding the compounds which act through inhibition of gonadotropin secretion (Dorfmann, 1971). To be maximally effective, an antiandrogen should be able to block the effect of androgens at the receptor level without exerting any androgenic activity by itself. The present data show that all progestins tested exert androgenic activity as assessed by the stimulation of growth of an androgen-sensitive tumor in an androgen-poor environment, namely female mice. Although an androgenic activity has been previously suggested for most of these compounds using androgen-sensitive parameters in normal tissues, it is likely that the present data obtained in vivo with androgen-sensitive cancer cells are even more relevant to human prostate cancer.

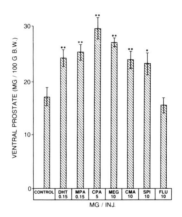

Fig. 4. Effect of treatment with the indicated doses of dihydrotestosterone (DHT), medroxyprogesterone acetate (MPA), cyproterone acetate (CPA), megestrol acetate (MEG), chlormadinone acetate (CMA), spironolactone (SPI) and Flutamide (FLU) on ventral prostate weight in the rat. The injections were performed twice daily for two weeks starting on the day of castration (*, p 0.05; **, p 0.01).

In agreement with the present fundamental data, it has recently been found in a study performed by the genitourinary tract cooperative group of the European Organization for Research on the Treatment of Cancer (EORTC) that the time to progression as well as survival were significantly shortened in the group of patients who received medroxyprogesterone acetate or compared to those who received diethylstilbestrol (Pavone-Macaluso et al., 1986). In fact, medroxyprogesterone acetate had a dramatic harmful effect on survival illustrated by a median life expectancy reduced to only one year as compared to approximately 2 years by all standard hormonal therapies reported to date. Such data illustrate the major importance of performing detailed preclinical studies of the properties of drugs to be used as antiandrogens for the treatment of androgen-sensitive diseases such as prostate cancer in order to identify the drugs having the most desirable activity profile.

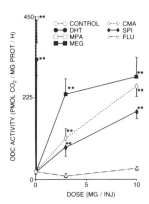

Fig. 5. Effect of increasing doses of medroxyprogesterone acetate (MPA), megestrol acetate (MEG), chlormadinone acetate (CMA), spironolactone (SPI), or Flutamide (FLU) on prostatic ornithine decarboxylase (ODC) activity in castrated rats. Subcutaneous injections were administered twice daily for 14 days starting on the day of castration. Results are presented as means ± SEM.

E. COMBINATION THERAPY IN PREVIOUSLY UNTREATED PATIENTS WITH CLINICAL STAGE D2 PROSTATE CANCER

Starting in March 1982, 199 patients with histology-proven prostatic adenocarcinoma and bone metastases visualized by bone scintigraphy (stage D2) took part in this multicentre study (Laval University Medical Center, Hôtel-Dieu Hospital, St-Jérôme and Hotel-Dieu Hospital, Lévis). All patients were entered into the study after written informed consent. All assays were done at the Laval University Medical Center. The mean duration of treatment was 771 days (88 to 1784). Last evaluation was on June 30th, 1987. The criteria for inclusion and exclusion were those of the US NPCP (Slack et al., 1984) except that a life expectancy of at least 90 days and normal blood counts were not used as criteria for exclusion. All patients presenting with stage D2 prostate cancer and having received no previous endocrine therapy or chemotherapy were thus included, the only exclusions being the presence of a second cancer (3 cases). The patients with very advanced disease with a short life expectancy were not excluded in order to more closely mimick the situation found in usual urological practice.

Demographic data and baseline profiles for patients who received the combination therapy (LUPCP, Laval University Prostate Cancer Program) and those of study 500 of the US NPCP who had orchiectomy or received DES are shown in Table 1. While the age of the patients, baseline pain and baseline performance status are not statistically different in the two studies, it can be seen that a larger number of patients in the present study had elevated serum prostatic acid phosphatase (PAP) values (p $<$ 0.01 versus NPCP 500), which is considered to be factor contributing toward a poorer prognosis. Fifty-two patients (26%) had a loss of body weight while fatigue and tiredness were present in 93 patients (47%). Neurological deficiency was found in 24 patients (12%). Hydronephrosis was seen in 18 patients (9%) while urinary obstruction was present in 8 patients (4%). Pain and abnormal performance were found in 126 (63%) and 86 (43%) patients, respectively (Tables 2 and 3). Location of metastases prior to combination therapy included bone metastases in all patients as well as involvement of lung (10, 5%), lymph nodes (32, 16%), bone marrow (5, 2.5%), brain (2, 1.0%) and liver (3, 1.5%).

TABLE 1

Demographic data and baseline profiles for patients of the studies using the combination therapy with [D-Trp^6] LHRH ethylamide and Flutamide (LUPCP) and ORCH/DES (study NPCP-500) [5].

	LUPCP	NPCP-500
Number of patients	199	83
Age (years)		
Mean	66	67
Median	65	66
Range	38–86	43–104
P-value (t test)		.91
Baseline Pain		
Present	126 (63%)	47 (57%)
Absent	73 (37%)	31 (37%)
Not specified	0	5 (6%)
P-value (Fisher's exact)		.22
Baseline Performance Status		
Normal	113 (57%)	37 (45%)
Symptomatic	53 (27%)	32 (39%)
Bedridden 50%	19 (9%)	8 (10%)
Bedridden 50%	10 (5%)	6 (7%)
Bedridden 100%	4 (2%)	0 (0%)
P-value (Mann Whitney)		.21
Elevated PAP (2ng/ml)		
Yes	177 (89%)	56 (67%)
No	22 (11%)	27 (33%)
P-value (Fisher's exact)		.01
Soft Tissue Metastases		
Yes	62 (31%)	24 (29%)
No	137 (69%)	59 (71%)
P-value (Fisher's exact)		.34

Of the 199 previously untreated stage D2 patients who had combination therapy, 189 received the combination treatment with the LHRH agonist [D-Trp^6, des-Gly-NH$_2$^{10}]LHRH ethylamide (Tryptex) in association with the pure antiandrogen 2-methyl-N-[4-nitro-3(trifluoromethyl)phenyl]propanamide (Flutamide, Euflex, Eulexin) while 10 had orchiectomy (instead of LHRH agonist treatment). No difference in the clinical response was observed between chemical or surgical castration. Twenty patients were originally started randomly

Table 2

Effect of combination therapy on the evolution of pain in previously untreated patients presenting with clinical stage D$_2$ prostate cancer (data are expressed in % of total number of patients).

PAIN	MONTHS OF TREATMENT								
	0	1	3	6	9	12	18	24	30
None	36.7	61.6	71	73.7	71.8	76.2	78.6	83.1	75.0
Moderate: permitting normal activity	35.1	34.3	26.5	23.7	25.9	16.4	17.5	10.4	18.5
Severe: interfering with daily activity and/or sleep	20.6	3.0	1.9	1.3	1.2	6.5	2.9	6.5	6.5
Very severe: constantly suffering from pain	5.0	1.0	0.6	1.3	1.2	0.8	0.9	0	0
Intolerable: requiring 100% bedridden	2.5	0	0	0	0	0	0	0	0

Table 3

Effect of combination therapy on the evolution of performance status in previously untreated patients presenting with clinical stage D_2 prostate cancer (data are expressed in % of total number).

PERFORMANCE	MONTHS OF TREATMENT								
	0	1	3	6	9	12	18	24	30
Normal	56.8	9.9	85.1	91.4	89.4	85.2	91.3	89.3	81.2
Ambulatory but symptomatic	26.6	23.3	14.3	5.3	4.7	7.8	4.8	8.0	8.3
Bedridden <50%	9.5	5.8	0.6	3.3	4.7	3.1	3.8	1.3	10.4
Bedridden >50%	5.0	0.9	0	0	1.2	3.9	0	1.3	0
Bedridden 100%	2.0	0	0	0	0	0	0	0	0

with the flutamide analog, 5,5-dimethyl-3-[4-nitro-3-(trifluromethyl)phenyl]-2,4-imidazolidione (RU23908, Anandron). However, the occurrence of visual side effects in 70% of the patients receiving Anandron led to an early change from Anandron to Flutamide and to the exclusive use of Flutamide in all patients since June 1983.

The LHRH agonist was injected subcutaneously at the daily dose of 500 µg at 0800h for 1 month followed by a 250 µg daily dose while Flutamide was given three times daily at 0700, 1500 and 2300h at the dose of 250 mg orally. The antiandrogen was started one day before first administration of the LHRH agonist or orchiectomy. Recent kinetic data (Simard et al., 1986) and information about the rapid changes of sensitivity of androgen-sensitive tumors when exposed to partial blockade of androgens (Labrie and Veilleux, 1986; Luthy and Labrie, 1987) indicate that the optimal time for first administration of Flutamide should be 2 hours before first injection of the LHRH agonist or orchiectomy. This schedule will be used routinely in the future. At the time of relapse under combination therapy, the treatment with Flutamide and [D-Trp^6, des-Gly-NH$_2^{10}$] LHRH ethylamide is continued. Moreover, in order to further block adrenal androgen secretion, aminoglutethimide is administered routinely at the dose of 250 mg every 8 hours in association with a low dose of hydrocortisone acetate (10 mg at 0700h, 5 mg at 1500h and 5 mg at 2100h). The tolerance to this additional therapy has generally been good.

Complete clinical, urological, biochemical and radiological evaluation of the patients was performed before starting treatment as described (Labrie et al., 1985a; Labrie et al., 1983). The initial evaluation included history, physical examination, bone scan, transrectal and transabdominal ultrasonography of the prostate, ultrasonography of the abdomen, chest roentgenogram and skeletal survey and sometimes computerized axial tomography (CAT) of the abdomen and pelvis as well as excretory urogram (IVP). Bone scans were evaluated by an independent group of radiologists unaware of the treatment of the patients. Performance status and pain were evaluated on a scale of 0 to 4 (see Tables 2 and 3). The follow-up was as described (Labrie et al., 1983), patients being evaluated at 1, 3, 6, 12 months and every 6th month thereafter.

Serum prostatic acid phosphatase (PAP) was measured by RIA as described (Labrie et al., 1983). The criteria of the U.S. National Prostatic Cancer Project, were used for assessment of objective response to treatment (Slack et al., 1984) (Table 4). Among the 186 patients evaluated in that study, one was lost to follow-up (he had a complete response when last seen), three decided on their own to stop therapy while they were responding (1 complete, 1 partial, and 1 stable) and treatment was interrupted in three other patients for intolerance (2 had diarrhea and 1 developped a pulmonary fibrosis). Statistical significance was measured according to the multiple-range test of Duncan-Kramer (Kramer, 1956), the Fisher's exact and the Mann Whitney tests (Armitage, 1971), when appropriate. The probabilities of continuing response and survival were calculated according to Kaplan and Meier (1958).

The efficacy of the combination therapy was assessed by analysis of: 1- best objective response measured according to the criteria of the US NPCP (Slack et al., 1984), (Table 4); 2- duration of objective response measured according to the same criteria and 3- survival. Since the results

Table 4

Objective Response to Therapy according to the National Prostatic Cancer Project (NPCP) Criteria (slightly modified)

Complete response (CR)

All of the following criteria:
1. Tumor masses, if present, totally disappeared and no new lesion appeared.
2. Elevated acid phosphatase, if present, returned to normal.
3. Osteolytic lesions, if present, recalcified.
4. Osteoblastic lesions, if present, disappeared, with a negative bone scan.
5. If hepatomegaly is a significant indicator, there must be a complete return in liver size to normal - i.e. no distention below both costal margins at the xiphoid process during quiet respiration without liver movement, and normalization of all pretreatment abnormalities of liver function, including bilirubin, SGOT and yGT.
6. No significant cancer-related deterioration in weight (10%), symptoms, or performance status.

Partial response (PR)

Any of the following criteria:
1. Recalcification of one or more of any osteolytic lesions.
2. A reduction by 50% in the number of increased uptake areas on the bone scan.
3. Decrease of 50% or more in cross-sectional area of any measurable lesion.
(except in the prostate which is not sufficient by itself).
4. If hepatomegaly is a significant indicator, there must be at least a 30% reduction in liver size indicated by a change in the measurements, and at least a 30% improvement of all pretreated abnormalities of liver function, including bilirubin, SGOT and yGT.

All of the following:
5. No new site of disease.
6. Acid phosphatase returned to normal.
7. No deterioration in weight (10%), symptoms, or performance status.

Objectively stable (S)

All of the following criteria:
1. No new lesion occurred and no measurable lesion increased more than 25% in cross-sectional area.
2. Elevated acid phosphatase, if present, decreased, though need not have returned to normal.
3. Osteolytic lesions, if present, did not appear to worsen.
4. Osteoblastic lesions, if present, remained stable on the bone scan (less than 25% increase in uptake)
5. Hepatomegaly, if present, did not appear to worsen by more than a 30% increase in the measurements, and signs of hepatic abnormalities did not worsen including bilirubin, SGOT and yGT.
6. No significant cancer-related deterioration in weight (10%), symptoms, or performance status.

Objective progression (P)

Any of the following criteria:
1. Significant cancer-related deterioration in weight (10%), symptoms, or performance status.
2. Appearance of new area of malignant disease by bone scan or x-ray or in soft tissue by other appropriate techniques.
3. Increase in any previously measurable lesion by greater than 25% in x-sectional area.
4. Development of recurring anemia, secondary to prostatic cancer.
5. Development of ureteral obstruction

NOTE: An increase in acid or alkaline phosphatase alone is not to be considered an indication of progression. These should be used in conjunction with other criteria

---

obtained using the same criteria in many recent studies on the blockade of testicular androgens achieved by various approaches have yielded almost superimposable results (Murphy et al., 1983; Smith et al., 1985; The Leuprolide Study Group, 1984); thus indicating the reliability of the criteria used, we have compared the present results with those obtained in those recent studies performed with comparable populations of patients who started treatment at the same stage of the disease.

As illustrated in Table 5 and Fig. 6, a positive objective response assessed according to the criteria of the US NPCP (Slack et al., 1984) has been obtained in 174 of 186 patients (93.5%), thus leaving only 12 patients (6.5%) with no response at the start of the treatment. The most striking effect is seen on complete responses (normal bone scan and no sign or symptom of prostate cancer) which have been observed in 49 of 186 patients (26.3%) as compared to an average of only 4.6% in the five recent studies limited to a blockade of testicular androgens (Murphy et al., 1983; Smith et al., 1985; The Leuprolide Study Group, 1984). The

Table 5

Best objective response (US NPCP criteria) achieved after combination therapy with FLutamide (LUPCP) compared to the results of orchiectomy or estrogens (NPCP-500, NPCP-1300, DES) and the LHRH agonist Leuprolide alone

Total evaluated	NPCP-500 N=83	NPCP-1300 N = 97	Leupro[a] lide-1 N = 47	Leupro[b] lide-2 N = 92	DES[b] N = 94	LUPCP N=186
Complete	10 (12%)	5 (5%)	1 (2%)	1 (1%)	2 (2%)	49 (26.3%)
Partial	24 (29%)	15 (15%)	18 (38%)	34 (37%)	41 (44%)	56 (30.1%)
Stable	33 (40%)	58 (60%)	15 (32%)	44 (48%)	37 (39%)	69 (37.1%)
Progression	16 (19%)	19 (20%)	13 (28%)	13 (14%)	14 (15%)	12 (6.5%)

a) Leuprolide 1: [28]
b) Leuprolide 2: [29]

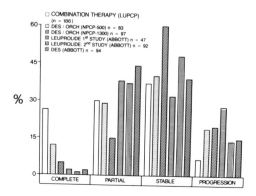

Fig. 6. Comparison of the best objective response rates (complete, partial, stable and progression) assessed according to the US NPCP criteria following combination therapy with Flutamide (LUPCP) and the five comparable studies using orchiectomy, DES or Leuprolide alone (Murphy et al., 1983; Smith et al., 1985; The Leuprolide Study Group, 1984). All were previously untreated patients with clinical stage D_2 prostate cancer.

---

rate of complete objective response is thus increased by 5.7-fold (p 0.01) (Fig. 6). The other striking finding illustrated in Table 5 and Fig. 6 is that only 6.5% of patients did not show an objective response at the start of the combination therapy while an average of 18% of patients failed to respond to monotherapy (orchiectomy, DES or Leuprolide alone) in the 5 other studies, thus representing a 3-fold difference in the percentage of non-responders or failure to treatment (p 0.01).

Figure 7 illustrates the probability of continuing response in the total group of 186 patients who could be evaluated. Quite remarquably, the probability of continuing response is 75.4% at 1 year (117 patients), 46.2% at 2 years (61 patients) and 36.6% at 3 years (23 patients). It can be seen in the same Figure that disease had progressed before 2 years in all patients treated with Leuprolide alone or DES (The Leuprolide Study Group, 1984).

It is also of interest to compare the probability of continuing response in patients who received the combination therapy (present study) and groups of similar patients who received DES, cyproterone acetate or medroxyprogesterone acetate (Pavone-Macaluso et al., 1986) (Fig. 8). Although the number of patients is smaller in the other groups, it can be seen that while the probability of continuing response after 1 year of treatment is 75.4% after combination therapy, it isreduced to 60, 33 and 23% after treatment with DES, cyproterone acetate and medroxyprogesterone acetate, respectively (Fig. 8).

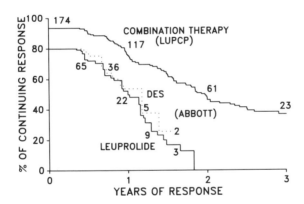

Fig. 7. Comparison of the probability of continuing response following combination therapy (this study, LUPCP) and the administration of Leuprolide alone or DES (The Leuprolide Study Group, 1984). The numbers on each curve correspond to the number of patients evaluated at that time period.

As expected from the 5.7-fold increase in the percentage of complete responders and the 3-fold decrease in the rate of non responders, survival after combination therapy is markedly improved during long-term treatment with the combination therapy (Fig. 9). In fact, following combination therapy, the probabilities of death after 2 and 4 years are 23.8 and 51.7%, respectively, as compared to 34 and 72%, 49 and 74% and 73 and 90% after treatment with DES, cyproterone acetate and medroxyprogesterone acetate, respectively (Pavone-Macaluso et al., 1986). The difference in death rate following combination therapy and the average results obtained by monotherapy, namely DES, cyproterone acetate and medroxyprogesterone acetate (EORTC study 30761) is better illustrated in Fig. 10. There is in fact an approximately 2-fold higher death rate up to 4 years of treatment in the EORTC study compared to our study using the combination therapy. It should be mentioned that the death rates presented include all causes of death, the death rate from causes other that prostatic cancer being approximately 5% per year in this age group.

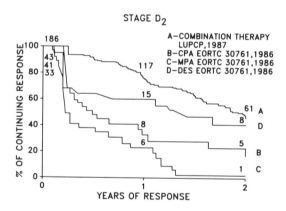

Fig. 8. Comparison of the probability of continuing response following combination therapy (this study, LUPCP) and the administration of DES, cyproterone acetate or medroxyprogesterone acetate (Pavone-Macaluso et al., 1986). The numbers on each curve correspond to the number of patients evaluated at that time period.

Since, in most other studies, the number of patients followed or alive after two years is usually too small, we have used 2 years as the time for comparison of survival rates achieved by different treatments. As illustrated in Fig. 11, while only 23.8% of patients were dead at 2 years of treatment with the combination therapy, an approximately 2-fold higher death rate ranging from 40.5 to 57.4% is found in the studies where patients were treated by monotherapy (DES, orchiectomy or LHRH agonist alone) (Jordan et al., 1977; Mettlin et al., 1982; Murphy et al., 1983; Smith et al., 1985; The Leuprolide Study Group, 1984; Labrie et al., 1985). As mentioned above, the death rates at 2 years in the EORTC study 30761 were 37, 49 and 73% in patients treated with DES, cyproterone acetate and medroxyprogesterone acetate, respectively (Pavone-Macaluso et al., 1986).

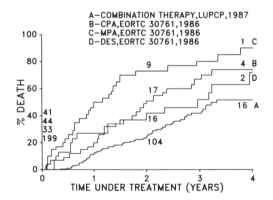

Fig. 9. Comparison of the probability of death following the combination therapy (A) or treatment with DES (D), cyproterone acetate (B) or medroxyprogesterone acetate (C) (Pavone-Macaluso et al., 1986). The numbers on each curve indicate the number of patients alive at that time period.

It is of interest to see in Fig. 12 that classification of patients according to the four categories of objective responses of the US NPCP has major prognostic value. While the small group of 12 non responders had an average 50% life expectancy of only 9.5 months, the best probability of survival is seen in those who had a complete response while less favorable prognoses are observed for the patients who showed partial and stable responses. In fact, at 3 years, the probabilities of survival are 95.9, 51.5 and 34.6%, respectively, in the categories of patients who have achieved complete, partial and stable responses as their best response to treatment (Fig. 12).

Almost all patients displayed various levels of prostatism which was improved during the first weeks of treatment. The rectal examination revealed an enlarged and hard prostate in 85% of cases. In all of them, the volume of the gland rapidly regressed and its consistency became normal during the first months of treatment. Rectal examination was confirmed by ultrasonography in most cases. Neurological deficiency and hydronephro sis were present in 12 and 9% of the patients, respectively, at the start of treatment. A total recovery from neurological symptoms was observed in 13 patients while hydronephrosis was corrected in all cases by treatment. Low urinary obstruction present in 8 patients disappeared without the need for TUR in any case.

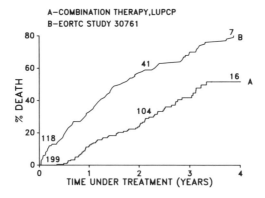

Fig. 10. Comparison of the probability of death following the combination therapy (A) and treatment with monotherapy (DES, cyproterone acetate or medroxyprogesterone acetate) (B). Curve B in this figure is the sum of curves B, C and D in fig. 4 (Pavone-Macaluso et al., 1986).

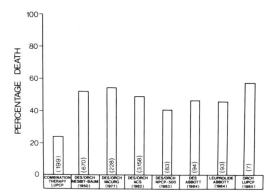

Fig. 11. Comparison of the death rate after 2 years of treatment with the combined androgen blockade (Laval University Prostate Cancer Program (LUPCP)) with results obtained with standard hormonal therapies (orchiectomy (ORCH) and/or estrogens or LHRH agonist alone) in previously untreated stage D patients: Nesbit and Baum's study (1950); study of the Veterans' Administration Cooperative Urology Research Group (VACURG) (1967); survey of the American College of Surgeons (ACS) (Mettlin et al., 1982); and study 500 of the US NPCP (Murphy et al., 1983); Leuprolide alone (1984) (Smith et al., 1985; The Leuprolide Study Group, 1984); and orchiectomy alone (LUPCP) (Labrie et al., 1985).

Performance status was rapidly and markedly improved. As shown in Table 3, eighty-six patients (43.2%) had an abnormal performance at start of treatment as compared to 15 and 9% 3 and 6 months later, respectively. In the 56 patients who had severe, very severe or intolerable pain before starting combination therapy, only 5 were still complaining of severe or very severe pain related to their cancer after 6 months, while none had intolerable pain compared to 5 at the beginning of treatment (Table 2).

At each visit, the patients answered a detailed questionnaire concerning any possible symptom or sign of intolerance to the drugs. Hot flashes were described spontaneously by the patients in approximately 50% of the cases after 1-3 months of treatment. Usually, the severity of the hot flashes decreased with time and disappeared within 2 years. A loss of libido was observed in approximately 75% of patients. However, it should be mentioned that in 25% of subjects, libido and potency were still present. No secondary effect which could not be attributed to hypoandrogenism was observed.

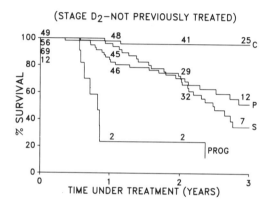

Fig. 12. Probability of survival according to the best objective response achieved (complete, partial, stable or progression) following combination therapy. The numbers indicate the numbers of patients evaluated at each time period.

The present data obtained in the largest study ever performed using the objective criteria of the US NPCP clearly indicate that the use of the combination therapy with Flutamide and castration (LHRH agonist or orchiectomy) significantly increases the rate of objective response, the duration of response as well as survival compared to the five comparable recent studies on the effect of monotherapy (orchiectomy, DES or Leuprolide alone).

The 5.7-fold increase in the rate of complete response is particularly striking. It should be mentioned that the 26.3% rate of complete response observed in the present

study includes patients recently entered into the study, a
situation which does not take into account the chance of
achieving a complete response by such patients at later time
intervals. In fact, for the 140 patients who have been under
combination therapy for 2 years, 31% of them have shown a
complete response as best response. Such a dramatic increase
in the rate of complete objective response has parallel
beneficial effects on the quality of life of these
patients.

As previously shown in study 500 of the US NPCP and
confirmed in the present study (Fig. 12), patients who show
a complete response have a much better probability of long
survival. In fact, it is well known that patients who do not
respond to endocrine therapy or continue to progress at the
start of treatment have an extremely poor prognosis, the
median life expectancy being approximately 6 months (Johnson
et al., 1977). The 3-fold decrease in the percentage of non
responders from 18 to 6.5% clearly illustrates the marked
advantages of the combination therapy upon both the quality
of life and survival. Since the rate of partial and stable
responses is not appreciably different between monotherapy
and dual combination therapy (Fig. 6), the major difference
between standard therapy and combination therapy is a shift
of patients from the category of non responders (short life
expectancy and poor quality of life) into that of complete
responders (increased life expectancy and good quality of
life).

F- COMBINATION THERAPY IN PREVIOUSLY UNTREATED STAGE C
PROSTATE CANCER PATIENTS

Since September 1982, seventy men with histology-proven
adenocarcinoma of the prostate were entered into this study
after written informed consent. Average age at entry in the
study was 69 years (from 48 to 88 years) with a median
follow-up of 666 days (62 to 1466). Complete clinical, urological, biochemical and radiological evaluation of the
patients was performed before starting treatment as described above.

As shown in Fig. 13, all except one evaluable patients
have shown a positive response to the treatment. Moreover,
only 5 patients have shown progression or treatment failure
to the combination therapy after an average treatment period

of 714 days. The probability of survival also illustrated in Fig. 13 is 98.2% and 93.7% at 1 and 2 years, respectively. Three patients died from prostate cancer while three died from other causes (myocardial infarct, pneumonia, and suicide). All three patients had been examined at our Prostate Cancer Clinic within 6 months of death and had been found to be clinically free of disease.

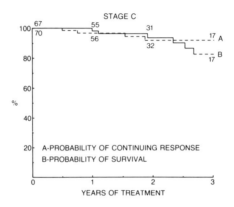

Fig. 13. Probabilities of continuing response (A) and survival (B) in patients with clinical stage C prostate cancer who received the combination therapy with Flutamide in association with the LHRH agonist D-Trp^6]LHRH ethylamide.

Serum prostatic acid phosphatase which was elevated in forty patients before starting treatment became normal in 44, 59, 91 and 96% of them, at respectively one, two, three and six months after the beginning of combination therapy. In all except one patient, the volume of the prostate rapidly regressed and its consistency became normal during the first nine months of treatment at rectal examination. These changes were confirmed by ultrasonography of the prostate in most patients. The low urinary tract obstruction present in 3 patients was corrected by treatment in all cases. In all these patients, the transurethral catether could be removed less than 21 days after the beginning of combination therapy. Hydronephrosis originally present in eight patients disappeared in all of them before six months of therapy.

The goal of therapy in stage C prostate cancer is local control of the tumor and prolongation of the interval free of disease (Gibbons et al., 1979; Tomlinson et al., 1977). The present data show that local control of the disease was achieved rapidly in all except one patient (98.6%). In the three patients who had a low urinary tract obstruction, the catheter could be removed within 3 weeks after starting treatment, thus indicating a rapid regression of the cancer at the prostatic level.

The present data suggest that administration of the combination therapy using a pure antiandrogen (Flutamide) in association with medical castration ([D-Trp^6]LHRH ethylamide) at the time of diagnosis of stage C prostate cancer has advantages over standard therapies and delayed treatment (Pavone-Macaluso et al., 1986; Paulson et al., 1978; Cupps et al., 1980; Smith et al., 1986). Since these recent studies have used comparable criteria of response, it seems appropriate and much informative to make a comparison with the results obtained in those studies. Since, in most studies, the number of patients at 3 years of treatment is too small, it seems more appropriate to use 2 years of treatment as the time of comparison.

As illustrated in Fig. 14, the probability of treatment failure at 2 years of treatment with the combination therapy is only 8.2% while 24 and 32% of patients have progressed to stage D2 after radiotherapy and delayed hormonal therapy, respectively (Paulson et al., 1978). Another study (Cupps et al., 1980) shows that 2 years after radiotherapy, the rate of progression to stage D2 is 18%. In a more recent study, 22% of stage C patients had progressed to stage D2 after 2 years of treatment with Estracyt or DES (Smith et al., 1986). In another recent study, the rate of progression to stage D2 at 2 years was 40, 34 and 66% after treatment with cyproterone acetate, DES and medroxyprogesterone acetate, respectively (Pavone-Macaluso et al., 1986). When all the above-mentioned data of monotherapy are combined (275 patients), the rate of treatment failure is on average 28.4%, a value 3.5 higher than that observed in the present study (8.2%).

Although progression to stage D2 is the early sign of treatment failure, it is of interest to compare, even at this early stage of the study, the survival rate so-far obtained under combination therapy with the results obtained in other studies using monotherapy. As illustrated in fig. 15, the death rate at 2 years following the start of combination therapy is 6.5%, this value being, on average, at 34% (5.2-fold difference) at the same time interval following treatment with DES or Estracyt (Smith et al., 1986). In the other EORTC study the death rates at 2 years after starting treatment with cyproterone acetate, DES or medroxyprogesterone acetate were 12, 22 and 31%, respectively (Pavone-Maculuso et al., 1986). When the above-mentioned data are pooled (513 patients), the average death rate at 2 years is 22.2% as compared to only 6.5% in the present study (3.4-fold difference).

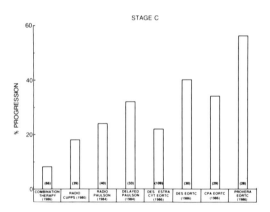

Fig. 14. Probability of continuing response in relapsing patients who show a positive response to combination therapy with Flutamide. The patients had clinical stage D2 prostate cancer and were in progression following standard endocrine therapy (orchiectomy, DES or LHRH agonist alone).

E- COMBINATION THERAPY IN PATIENTS RELAPSING AFTER STANDARD ENDOCRINE THERAPY

All patients with biopsy-proven stage D_2 prostatic adenocarcinoma showing disease progression after orchiectomy, diethylstilboestrol (DES) or an LHRH agonist alone were included in the study. Flutamide (250 mg every eight hours) was given alone in castrated patients. In patients previously treated with DES, Flutamide was given in combination with the LHRH agonist [D-Trp^6]-LHRH ethylamide (LHRH-A) administered at the dose of 500 ug, s.c., daily for the first month followed by the daily dose of 250 ug s.c., thereafter. For those already receiving an LHRH agonist, Flutamide was added to the treatment. Evaluation of response was performed according to the objective criteria of the National Prostatic Cancer Project (NPCP) (Slack et al., 1984).

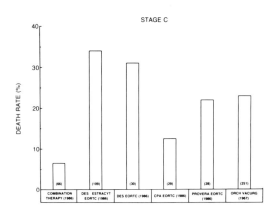

Fig. 15. Comparison of the probability of survival following combination therapy with Flutamide and standard therapy (chemotherapy and/or DES was added to orchiectomy and viceversa).

Of the 209 patients who took part in the study, one hundred and sixty-six patients (79%) complained of pain, one hundred fifty-three (73%) had an elevated serum prostatic acid phosphatase (PAP) and thirty (14.4%) displayed hydronephrosis at ultrasonography of the abdomen and intravenous pyelography. Forty percent of the patients were symptomatic but ambulant (84 of 209). Twenty additional patients (9.6%) were bedridden less than 50% of the time and 23 (11%) were bedridden greater than 50% of the time while 3 (1.4%) patients were bedridden 100% of the time. Forty-three percent of the patients (90/209) complained of anorexia and weight loss prior to therapy (Table 6).

Location of metastases prior to the combination therapy included bone metastases in all patients as well as involvement of the lungs (22, 10.5%), distant and pelvic lymph nodes (10, 4.8%), liver (3, 1.4%) and bone marrow (12, 5.7%). Previous therapy included orchiectomy in 116 patients (55.5%). For the patients who had received DES or an LHRH agonist alone, pretreatment levels of testosterone were in the orchiectomized range. Fifty patients (24%) had received at least one course of radiotherapy for metastatic disease and none of them had received chemotherapy. The probability of survival was calculated according to Kaplan and Meier (1958).

Table 6

Symptoms and performance status in 209 patients with stage D prostate cancer in relapse after castration

Symptoms	Number of patients	Percentage
Loss of appetite	59	28.2
Loss of body weight	76	36.4
Pain	166	79.0
Hydronephrosis	30	14.4
Urinary obstruction	2	0.9
Performance		
Bedridden 100%	3	1.4
Bedridden > 50%	23	11.0
Bedridden < 50%	20	9.6
Symptomatic	84	40.0
Normal	79	37.8

Table 7

Effect of combination therapy with Flutamide on the best objective response assessed according to the criteria of the US NPCP (Slack et al., 1984) in 209 patients relapsing after castration (surgical or medical)

DAYS OF TREATMENT MEAN (LIMITS)	BEST OBJECTIVE RESPONSE			
	COMPLETE	PARTIAL	STABLE	PROGRESSION
426 (61-1480)	13 6.2%	20 9.6%	39 18.6%	137 65.5%

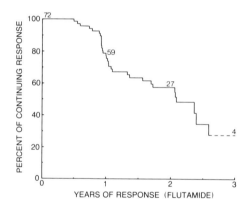

Fig. 16. Comparison of the probability of survival in responders and non responders following combination therapy with Flutamide in patients with clinical stage D2 prostate cancer in progression after standard endocrine therapy (orchiectomy, DES or LHRH agonist alone).

RESULTS

As shown in Table 7, thirteen (6.2%) patients had a complete response to the treatment, twenty (9.6%) showed a partial response, thirty-nine (18.7%) obtained an objective stable response, while one hundred thirty-seven (65.5%) patients continued to progress. It can be seen in Fig. 16 that the median duration of positive response after combination therapy is 24 months. In addition to the responders, an improvement in pain and performance was observed in an additional 30% of patients for 1 to 3 months. In the responders, the improvement in pain and performance usually lasted a few months more than the objective response illustrated in Fig 16.

Figure 17 shows the probability of survival in patients in relapse treated by the combination therapy (curve A) compared to the survival observed in a similar group of

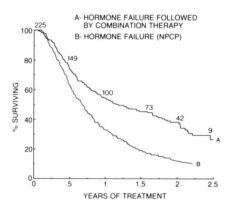

Fig. 17. Probability of survival according to the best objective response achieved (complete, partial, stable and progression) following combination therapy with Flutamide. The calculations were made starting at the time of treatment with combination therapy. Numbers on the curve represent the number of patients assessed or non-censored at each period of time.

patients treated by conventional therapy (NPCP) (Priore, 1984). The probability of survival at 2 years after starting the combination therapy is 38% compared to 16% in a similar group of patients who had progressed to hormonal therapy (ORCH/DES) and received chemotherapy or DES following surgical castration and vice-versa (Priore, 1984). Figure 18 shows that while the median life expectancy is 8.1 months in the non responders, it is increased to more than 2.5 years in the group of responders.

It is of interest to see in Fig. 19 the survival rates calculated for each group of objective responses observed after combination therapy. All the patients who have reached a complete objective response are still alive. On the other hand, the patients who have reached partial and stable responses show probabilities of survival at 2 years of 87 and

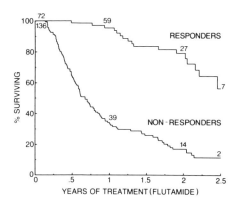

Fig. 18. Comparison of the probability of treatment failure (progression) in patients with stage C prostate cancer who received the combinatio ntherapy with Flutamide and the LHRH agonist [D-Trp^6]LHRH ethylamide (present data), radiotherapy (Paulson, 1984; Cupps et al., 1980), delayed treatment (Paulson, 1984), DES/Estracyt (Smith et al., 1986), DES (Pavone-Macaluso et al., 1986), cyproterone acetate (CPA) (Pavone-Macaluso et al., 1986) or medroxyprogesterone acetate (Provera) (Pavone-Macaluso et al., 1986).

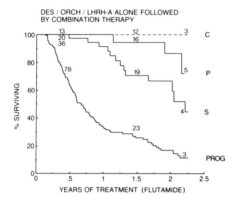

Fig. 19. Comparison of the probability of death in patients with clinical stage C prostate cancer who received the combination therapy with Flutamide and the LHRH agonist [D-Trp[6]]LHRH ethylamide (present data), DES/Estracyt (Smith et al., 1986), DES (Pavone-Macaluso et al., 1986), cyproterone acetate (CPA) (Pavone-Macaluso et al., 1986), medroxyprogesterone acetate (Provera) (Pavone-Macaluso et al., 1986) or orchiectomy (Veterans Administration Cooperative Urological Research Group, 1967).

67%, respectively, while continuing progression of the disease (or failure to respond objectively) leads to a 17% survival rate at the same time interval.

In agreement with our previous results obtained in previously untreated patients, the combination therapy with Flutamide causes a 26 to 49% decrease in the serum levels of the four adrenal steroids dehydroepiandrosterone sulfate, dehydroepiandrosterone, androst-5-ene-3β,17β-diol and androstenedione while the serum concentration of cortisol remained unchanged. Hot flashes were recorded in about 80% of patients while gastrointestinal side effects (nausea, heartburn and diarrhea) were seen in about 8% of the patients. None of these required an arrest of Flutamide treatment.

The present data clearly demonstrate in the largest study ever performed in a group of patients relapsing after castration that the simple addition of Flutamide blocks progression of the disease in as many as one third of them with no or minimal side effects. The finding of complete, partial and stable disease in respectively 6.2, 9.6 and 18.7% of the patients illustrates the important influence of the treatment on the quality of life of these patients. Moreover, the median life expectancy is increased from 8.1 months in non responders to an average of 36 months in responders ($p < 0.01$). It is quite remarquable that all signs of bone metastases and other signs and symptoms of prostate cancer have completely disappeared in 13 patients (6.2%). No death has in fact occurred in the patients who have shown a complete objective response assessed according to the criteria of the US NPCP.

It is of interest to mention that the three categories of positive objective responses, namely complete, partial and stable are directly correlated with the survival rate (Fig 19), the best responders having the best prognosis for survival while those who are not responding to the combination therapy show the worst prognosis. Patients who had a stable response have a better prognosis of survival than the non responders, but less than those who reached partial or complete responses.

Clinical evidence for a role of adrenal androgens in prostate cancer is most convincing. Several studies indicate that the remission rate observed following hypophysectomy in patients who relapse following previous endocrine therapy is 50% or more (Scott and Schirmer, 1966; Luft and Olivecrona, 1957; Straffon et al., 1968; Ferguson and Phillips, 1982). Although the duration of remission was usually short, the benefits were highly significant. In the early studies on the effect of hypophysectomy, the number of patients was generally small due, to a large extent, to the morbidity and the high mortality rate of craniotomy in elderly and usually debilitated patients. More recently, in a series of 19 patients who had cryohypophysectomy after relapsing from previous endocrine therapy, an objective response was seen in 7 (37%) patients while subjective improvement was seen in 5 (26%) other cases, for a total improvement in 63% of patients. The mean duration of survival in those who responded was 11.5 months as compared to 3.8 months in those who did not respond (Maddy et al., 1971).

The common belief that patients in relapse after castration or treatment with estrogens have exclusively "androgen-insensitive" tumors should be abandonned. In fact, it is most likely that androgen-sensitive tumors are present at all stages of prostate cancer in all patients and that maximal androgen blockade should be performed in all cases. Instead of being "androgen-insensitive", most of (if not all) the tumors which continue to grow after castration are androgen- hypersensitive. These tumors are able to grow in the presence of the "low" level of androgens of adrenal origin left after castration. Control of the growth of these tumors requires further androgen blockade. This affirmation is supported by convincing clinical data as well as by well-established fundamental observations (Labrie and Veilleux, 1986, Labrie et al., 1987a).

As mentioned above, Flutamide should always be given in association with castration and should never be administered alone. The reason not to use Flutamide alone is that when given alone, the levels of endogenous DHT in the prostate cancer tissue remain too high since the antiandrogen is unable to reduce DHT binding to the androgen receptor to sufficiently low levels. Following castration, 50% of DHT is automatically removed, thus markedly facilitating the competitive action of the antiandrogen. Furthermore, since the combination therapy with Flutamide reduces serum levels of precursor adrenal androgens by another 50% (Labrie et al., 1985a), it follows that intraprostatic DHT levels are further reduced to approximately 25% of control during combination therapy, thus greatly facilitating the blockade of the androgen receptor by Flutamide.

Using the known affinities of DHT and OH-Flutamide (active metabolite of Flutamide) and the intraprostatic concentration of the two substances (Simard et al., 1986; Labrie et al., 1987a), we have calculated that Flutamide administered alone leaves approximately 40% of DHT free to bind to the androgen receptor in the prostatic tissue. However, when the antiandrogen is administered in castrated patients as part of the combination therapy, less than 5% of DHT is left free to bind to the androgen receptor (Labrie et al., 1987a). Direct proof of the ability of Flutamide to displace DHT from the androgen receptor has been provided by our recent data showing that the intraprostatic DHT concentration decreases from 4.2 ± 0.6 ng/g tissue in untreated men to less than 0.2 ng/g tissue during combination therapy (4.8% of control) (Labrie et al., 1987a).

The present data clearly indicate that Flutamide in association with castration (surgical or medical) is the drug of choice to be used in patients relapsing after castration or treatment with an estrogen or an LHRH agonist alone. Important subjective and objective benefits including increased survival in the responders are thus easily obtained with no significant side effect.

In addition to the long-term benefitial effects of combination therapy, the use of Flutamide at the start of treatment eliminates the unnecessary risks of disease flare wich are known to occur in a significant proportion of patients treated with an LHRH agonist alone (Nesbit and Baum, 1950; Kahan et al., 1984; Waxman et al., 1983). It seems obvious that exposure of the tumor cells to supraphysiological levels of androgens represents an increased stimulus for the tumor to grow and to metastasize. The present data clearly show that the pure antiandrogen permits to take advantage of the well tolerated LHRH agonists as substitutes for orchiectomy and estrogens by eliminating the risk of disease flare (Labrie et al., 1987b).

The most likely explanation for the difference observed between the present results and those of previous studies is that previous hormonal therapy was limited to the neutralization of androgens of testicular origin by surgical castration and/or estrogens while the present approach achieves more complete blockade of androgens of both testicular and adrenal origin at the start of treatment. As mentioned above, a large number of reports have shown that neutralization of adrenal androgens has beneficial effects on prostate cancer (Labrie et al., 1985a; Murray and Pitt, 1985; Maddy et al., 1971). However, in the past, medical or surgical adrenalectomy or hypophysectomy was never performed as a first approach in combination with blockade of testicular androgens. The neutralization of adrenal androgens was always achieved as a second step following the lack of response to castration or when relapse of the disease had occurred after a period of remission (Labrie et al., 1985a).

An unexpected but most important additional benefit of combined antiandrogen treatment is that it inhibits by approx 50% the serum levels of adrenal steroids responsible for the formation of active androgens in prostatic cancer tissue, especially dehydroepiandrosterone (DHEA), DHEA-sulfate (DHEA-S), androstenedione and androst-5- ene-3β,17β-diol (Labrie et al., 1985a; Bélanger et al., 1986; Labrie et al., 1986; Bélanger et al., 1984). This approximately 50% decrease in the serum levels of precursor steroids should lead to a similar decrease in the level of active androgens in the prostatic cancer, thus decreasing the stimulatory androgenic influence on cancer growth. This decrease in the local intraprostatic concentration of DHT should facilitate the inhibitory action of the antiandrogen. It is thus quite remarkable that the combination treatment, in addition to completely blocking testicular androgen secretion as well as inhibiting the peripheral action of androgens, can also achieve a partial medical adrenalectomy selective for androgen precursors and not affecting the secretion of cortisol (Labrie et al., 1985a; Bélanger et al., 1984).

If one accepts that androgens stimulate prostate cancer growth, the next logical step in the treatment of this disease is to eliminate, as much as possible, all androgenic influences on prostate cancer with the best available drugs. Since the testes represent approximately 50% of androgens and appropriate means are readily available to eliminate this source of androgens, this should be an essential component of any antihormonal therapy of prostate cancer. With to-day's knowledge, the choice is between orchiectomy and the use of LHRH agonists. For the patients who accept surgical castration, this is certainly most valid. However, LHRH agonists are now widely available and there is no doubt that these peptides are a well-tolerated, safe and efficient way to achieve a complete blockade of testicular androgen secretion (Labrie et al., 1980; Labrie et al., 1985a,b; Labrie et al., 1986).

Due to their high rate of serious cardiovascular side effects, estrogens, in our opinion, are no longer justified. In addition to a death rate as high as 15% due to estrogens during the first year of treatment (Glashan and Robinson, 1981), a finding which is sufficient by itself to prohibit their use, there is also evidence that estrogens can increase the level of prostatic androgen receptors, thus increasing the local activity of androgens (Moore et al.,

1979; Mobbs et al., 1983). The elimination of approximately 50% of the androgens active in prostate cancer can thus be easily achieved by orchiectomy or treatment with LHRH agonists without any side-effect other than those related to the blockade of testicular androgens, namely hot flashes and a decrease or loss of libido.

Although more complete inhibition of androgens remains a possibility, the results obtained using the combined use of an LHRH agonist and a pure antiandrogen already show marked advantages over previous therapies limited to partial neutralization of androgens. It should be added that the above- described advantages of combination therapy are obtained without additional side effects, thus permitting extra months or years with a good quality of life.

Our previous data have indicated the advantages of combination therapy in patients with stage D2 prostate cancer (Labrie et al., 1985a,c; Labrie et al., 1986). The present data clearly suggest that the combination therapy is even more advantageous when applied earlier in the disease. The recent experimental evidence indicating the development of a marked heterogeneity of the sensitivity to androgens in tumors in vivo as well as in vitro in cells in culture (Labrie and Veilleux, 1986) provides strong support for the present clinical findings. We have in fact found that a marked heterogeneity of the responsiveness to androgens develops in vivo in the mouse androgen-sensitive Shionogi tumor some cells being up to 1200 times more sensitive than others to the androgen DHT. Moreover, when a clone originating from a single cell was grown in vitro under defined conditions, the same heterogeneity of sensitivity to DHT developped rapidly (Labrie and Veilleux, 1986).

Since every cancer cell division carries the risk of additional hete rogeneity, it seems logical to expect a less favorable efficacy when the combined antiandrogen treatment is applied at a later stage of the disease. As much as early tumors are more likely to be composed of cells which resemble the original normal prostatic cell and thus respond well to androgen blockade, it is expected that more advanced prostate cancer contains more clones of cells which have deviated from the normal pheno- type and have acquired a different sensitivity to androgens (both hypersensitive and hyposensitive). In analogy with the experimental Shionogi tumor, the prostatic tumors having a high sensitivity to androgens will be difficult to block since such high sensitivity permits them to grow on the small amounts of DHT left free under antiandrogen treatment.

Although, as mentioned earlier, the present study is non randomized and the possibility of a bias in the patient population exists, the exceptionally large difference observed when comparing to all available studies is highly suggestive of the advantages of the combination therapy over previous approaches. It should be mentioned that our population of patients is, if something, less favorable since it includes stage C, D_0 and D_1 patients. In fact, due to the unavoidable limitations of the available staging techniques, no routine lymphography or staging pelvic lymphadenectomy was performed to differentiate between stages C and D_1. As shown by the finding of elevated serum PAP levels in 40 of 70 patients (57%), a maximum of 33% of our patient population was still at stage C while 57% were at stage D_0 or D_1.

In agreement with previous suggestions (Grayhack and Kozlowski, 1980), the present data provide strong support for early treatment of prostatic cancer by endocrine therapy. The highly probable arguments which, in our opinion, strongly support early combination therapy are the following: 1) untreated stage C or D prostate cancer is usually progressive; b) minimizing tumor load is likely to facilitate individual immune antitumor defenses: c) minimizing cancer cell division is likely to decrease the appearance of more undifferentiated cell clones resistant to antiandrogen blockade; d) progressive disease is more likely to endanger the quality of life than the antiandrogen treatment; e) early recognition of non or poor responders to available antiandrogen blockade permits initiation of other therapies before deterioration of the general status of the patient.

When treatment is delayed until the disease has become symptomatic, the patient has to recover from one or more of the following complications: bone pain, anorexia, weight loss, urinary obstruction, cord compression and/or anemia. The argument for early administration of combination therapy is particularly strong when one considers the relative lack of undesirable side effects of the treatment. In fact, in patients who do not accept orchiectomy, the possibility of achieving medical castration with LHRH agonists avoids all the cardio-vascular problems associated with the use of estrogens (Glashan and Robinson, 1981; Veterans Administration Cooperative Urological Research Group, 1967; Hedlund et al., 1980) while the pure antiandrogen used in this study is devoid of significant side-effects. Hot flashes and a decreased or loss of potency and libido are usually found in patients receiving the combination therapy but it should be mentioned that these side-effects essentially due to the blockade of androgen action.are not new and were also present with previous therapies.

REFERENCES

Armitage T (1971). Statistical methods in Medical Research, Blackwell Scintific Publications, Oxford.
Bélanger A, Dupont A, Labrie F (1984). Inhibition of basal and adrenocorticotropin-stimulated plasma levels of adrenal androgens after treatment with an antiandrogen in castrated patients with prostatic cancer. J Clin Endocrinol Metab 59:422-426.
Bélanger A, Brochu M, Cliche J (1986). Levels of plasma steroid glucuronides in intact and castrated men with prostatic cacner. J Clin Endocrinol Metab 62:812-815.
Cupps RE, Utz DC, Fleming TR, Carson CC, Zincke H, Myers RP (1980). Definitive radiation therapy for prostatic carcinoma: Mayo Clinic Experience. J Urol 124:855-859.
Dexter D, Calabressi P (1982). Intraneoplastic diversity. Biochim Biophys Acta 694:97-112.
Dorfmann RI (1971). Antiandrogens. In Jarnes V, Martini L (eds): "Proc of the Third Int Congress on Hormonal Steroids", Excerpta Medica, Amsterdam, p. 995.
Farnsworth WE, Brown JR (1976). Androgen of the human prostate. Endocr Res Commun 3:105-117.
Faure N, Labrie F, Lemay A, Bélanger A, Gourdeau Y, Laroche B, Robert G (1982). Inhibition of serum androgen levels by chronic intranasal and subcutaneous administration of a potent luteinizing hormone-releasing hormone (GnRH) agonist in adult men. Fertil Steril 37:416-424.
Ferguson JD (1975). Limits and indication for adrenalectomy and hypophysectomy in the treatment of prostatic cancer. In Bracci U, Di Silverio F (eds): "Hormonal Therapy of Prostatic Cancer", Perlermo, Cofese Edizioni, pp. 201-207,

Fowler JE, Whitmore WF Jr (1981). The response of metastatic adenocarcinoma of the prostate of exogenous testosterone. J Urol 126:372-375.
Geller J, Albert JD, Nachtsheim DA, Loza DC (1984). Comparison of prostatic cancer tissue dehydrotestosterone levels at the time of relapse following orchiectomy or estrogen therapy. J Urol 132:693-696.
Gibbons RP, Mason JT, Correa RJ Jr, Cummings KB, Taylor WJ, Hafermann MD, Richardson RG (1979). Carcinoma of the prostate: local control with external beam radiation therapy. J Urol 121:310-312.
Glashan RW, Robinson MRG (1981). Cardiovascular complications in the treatment of prostatic carcinoma. Br J Urol 53:624-626.
Graf KJ, Kleinechke RI, Neumann F (1974). The stimulation of male duct derivatives in female guinea pigs with an antiandrogen cyproterone acetate. J Reprod Fertil 39:311-317.
Grayhack JT, Kozlowski JM (1980). Endocrine therapy in the management of advanced prostatic cancer: the case for early initiation of treatment. Urol Clin North Am 7:639-643.
Hamada H, Neumann F, Junkmann K (1963). Intrauterine antimaskuline beeinflussing von rattenfetendurch ein stark gestagen wirksames steroid. Steroid Acta Endocrinol 44:330-338.
Hedlund PO, Gustafsson H, Sjogren S (1980). Cardiovascular complications to treatment of prostatic cancer with estramustine phosphate (Estracyt) or conventional oestrogen. A follow-up of 212 randomized patients. Scand J Urol Nephrol [Suppl] 55:103-105.
Huggins C, Hodges CV (1941). Studies of prostatic cancer. I. Effect of castration, estrogen and androgen injections on serum phosphatases in metastatic carcinoma of the prostate. Cancer Res 1:293-297.
Huggins C (1947). Antiandrogenic treatment of prostatic carcinoma in man, approaches to tumor chemotherapy. Am Ass Adv Sci 379-383.
Huggins C, Scott WW (1945). Bileteral adrenalectomy in prostatic cancer. Ann Surg 122:1031-1041.
Johnson DE, Kaesler KE, Ayala AG (1981). Megestrol acetate for treatment of prostatic carcinoma. Br J urol 53: 624-630.

Johnson DE, Scott WW, Gibbons RP, Prout GR, Schmidt JD, Chu JTM, Gaeta J, Sarott J, Murphy GP (1977). National randomized study of chemotherapeutic agents in advanced prostatic carcinoma: progress report. Cancer Treat Rep 61:317-323.
Jordan WP Jr, Blackard CE, Byar DP (1977). Reconsideration of orchiectomy in the treatment of advanced prostatic carcinoma. South Med J 70:1411-1413.
Kahan A, Delrieu F, Amor B, Chiche R, Steg A (1984). Disease flare induced by D-Trp^6-GnRH analogue in patients with metastatic prostatic cancer. Lancet 1:971.
Kaplan EL, Meier P (1958). Non parametric estimation from incomplete observations. Am Stat Ass J 53:457-481.
Kramer CY (1956). Extension of multiple-range tests to groups means with unique numbers of replications. Biometrics 12: 307-310.
Labrie F (1984). A new approach in the hormonal treatment of prostate cancer: complete instead of partial blockade of androgens. Int J Androl 7:1-4.
Labrie F, Auclair C, Cusan L, Kelly PA, Pelletier G, Ferland L (1978). Inhibitory effects of LHRH and its agonists on testicular gonadotropin receptors and spermatogenesis in the rat. In Hansson V (ed) "Endocrine approach to Male Contraception" Int J Androl (suppl 2):303-308.
Labrie F, Bélanger A, Cusan L, Séguin C, Pelletier G, Kelly PA, Lefebvre FA, Lemay A, Raynaud JP (1980). Antifertility effects of LHRH agonists in the male. J Androl 1:209-228.
Labrie F, Dupont A, Bélanger A, Cusan L, Lacourcière Y, Monfette G, Laberge JG, Emond JP, Fazekas ATA, Raynaud JP, Husson JM (1982). New hormonal therapy in prostatic carcinoma: combined treatment with an LHRH agonist and an antiandrogen. Clin Invest Med 5:267-275.
Labrie F, Dupont A, Bélanger A, Lachance R, Giguère M (1985a). Long-term treatment with luteinizing hormone-releasing hormone agonists and maintenance of serum testosterone to castration concentrations. Brit Med J 291:369-370.
Labrie F, Dupont A, Bélanger A (1985b). Complete androgen blockade for the treatment of prostate cancer. In De Vita VT, Hellman S, Rosenberg SA, (eds) "Important Advances in Oncology", J.B. Lippincott, Philadelphia, pp. 193-217.

Labrie F, Dupont A, Bélanger A, Giguère M, Lacourcière Y, Emond J, Monfette G, Bergeron V (1985c). Combination therapy with Flutamide and castration (LHRH agonist or orchiectomy) in advanced prostate cancer: a marked improvement in response and survival. J Steroid Biochem 23:833-841.

Labrie F, Veilleux R (1986). A wide range of sensitivities to androgens develops in cloned Shionogi mouse mammary tumor cells. Prostate 8:293-300.

Labrie F, Dupont A, Bélanger A, St-Arnaud R, Giguère M, Lacourcière Y, Emond J, Monfette G (1986). Treatment of prostate cancer with gonadotroin-releasing hormone agonists. Endocr Rev 7:67-74.

Labrie F, Dupont A, Bélanger A, Lacourcière Y, Raynaud JP, Husson JM, Gareau J, FAzekas ATA, Sandow J, Monfette G, Girard JG, Emond J, Houle JG (1983). New approach in the treatment of prostate cancer: complete instead of partial withdrawal of androgens. Prostate 4:579-594.

Labrie F, Luthy I, Veilleux R, Simard J, Bélanger A, Dupont A (1987a). New concepts on the androgen sensitivity of prostate cancer. In Murphy G. (ed) "Second Int Symposium on Prostate Cancer", New York: Alan R. Liss Inc., in press.

Labrie F, Dupont A, Bélanger A, Lachance R (1987b). Flutamide eliminates the risk of disease flare in prostatic patients treated with a luteinizing hormone-releasing hormone agonist. J Urology, in press.

Luft, R., Olivecrona H (1957). Hypophysectomy in the management of neoplastic disease. Bull NY Acad Med 33:5-16.

Luthy I, Labrie F (1987). Development of Androgen Resistance in Mouse Mammary Tumor Cells Can Be Prevented by the Antiandrogen Flutamide. Prostate 10:89-94.

Maddy JA, Winternitz WW, Norrell H (1971). Cryohypophysectomy in the management of advanced prostatic cancer. Cancer 28:322-328.

Mettlin C, Natarajan N, Murphy GP (1982). Recent patterns of care of prostatic cancer patients in the United States: results from the surveys of the American College of Surgeons Commission on Cancer. Int Adv Surg Oncol 5:277-321.

Mobbs BG, Johnson IE, Connolly JG, Thompson J (1983). Concentration and cellular distribution of androgen receptor in human prostatic neopolasia: can estrogen treatment increase androgen receptor content? J Steroid Biochem 19:1279-1290.

Moore RJ, Gazak JM, Wilson JD (1979). Regulation of cytoplasmic dihydrotestosterone binding in dog prostate by 17β-estradiol. J Clin Invest 63:351-357.
Mowszowicz I, Bieber DE, Chung KW, Bullock LP, Bardin CW (1974). Synandrogenic and antiandrogenic effect of progestins: comparison with non-progestational antiandrogens. Endocrinology 95:1589-1599.
Murray R, Pitt P (1985). Treatment of advanced prostatic cancer resistant to conventional therapy with aminoglutethimide. Eur J Cancer Clin Oncol 21:453-458.
Murphy GP, Beckley S, Brady MF, Chu M, DeKernion JB, Dhabuwala C, Gaeta JF, Gibbons RP, Loening S, McKiel CF, McLeod DG, Pontes JE, Prout Gr, Scardino PT, Schlegel JU, Schmidt JD, Scott WW, Slack NH, Soloway M (1983). Treatment of newly diagnosed metastatic prostate cancer patients with chemotherapy agents in combination with hormones versus hormones alone. Cancer 51:1264-1272.
Neri R, Monahan MD, Meyer JG, Afonso BA, Tachnick IA (1967). Biological studies on an antiandrogen (SH-714). Eur J Pharmacol 1:438-444.
Nesbit RM, Baum W (1950). Endocrine control of prostatic carcinoma: clinical and statistical survey of 1818 cases. JAMA 143:1317-1320.
Noguchi S, Nishizama Y, Uchida N, Yamaguchi K, Sato B, Kitamura Y, Matsumoto K (1985). Stimulative effect of physiological doses of androgens or pharmacological doses of estrogen on growth of Shionogi carcinoma 115 in mice. Cancer Res 45:5746-5750.
Paulson DF (1978).The role of endocrine therapy in the management of prostate cancer. In Skinner DG and De Kernion JB (eds). "Genitourinary Cancer" W.B. Saunders, Philadelphia.
Pavone-Macaluso M, De Voogt HJ, Viggiano G, Barasolo E, Hardennois B, De Pauw M, Silvester R (1986). Comparison of diethylstilbestrol, cyproterone acetate and medroxyprogesterone acetate in the treatment of advanced prostatic cancer: final analysis of the randomized phase III trial of the European Organization for Research on Treatment of Cancer Urological Group. J Urol 1:624-631.
Poyet P, Labrie F (1985). Comparison of the antiandrogenic/androgenic activities of flutamide, cyproterone acetate and megestrol acetate. Mol Cell Endocrinol 32:283-288.
Priore R (1984). Prognostic factors in NPCP protocols. In National Prostatic Cancer Project, Treatment Subgroup Meeting, Minutes, New Orleans.

Resnick MI, Grayhack JT (1975). Treatment of stage IV carcinoma of the prostate. Urol Clin North Am 2:141-161.

Robinson MR, Shearer RJ, Fergusson JD (1974). Adrenal suppression in the treatment of carcinoma of the prostate. Brit J Urol 46:555-559.

Sanford EJ, Paulson DF, Rohner TJ, Drago JR, Santen RJ, Bardin CW (1977). The effects of castration on adrenal testosterone secretion in men with prostatic carcinoma. J Urol 118:1019-1021.

Silverberg E, Lubera JA (1983). A review of American Cancer Society estimates of cancer cases and deaths. CA 33:2-25.

Simard, J, Luthy I, Guay J, Bélanger A, Labrie F (1986). Characteristics of interaction of the antiandrogen flutamide with the androgen receptor in various target tissues. Mol Cell Endocrinol 44:261-270.

Slack NH, Murphy GD, NPCP participants (1984). Criteria for evaluating patient responses to treatment modalities for prostatic cancer. Urol Clin North Am 11:337-342.

Smith JA, Glode LM, Wettlaufer JN, Stein BS, Glass AG, Max DT, Anbar D, Jagst CL, Murphy GP (1985). Clinical effects of gonadotropin-releasing hormone analogue in metastatic carcinoma of the prostate. Urology 20:106-112.

Smith PH, Suciu S, Robinson MRG, Richards B, Bastable JRG, Glasdhan RW, Bouffioux C, Lardennois B, Williams RE, de Pauw M, Sylvester R (1986). A comparison of the effect of diethylstilbestrol with low dose estramustine phosphate in the treatment of advanced prostatic cancer: final analysis of a phase III trial of the European Organization for Research on Treatment of Cancer. J Urol 136:619-623.

Sogani PC, Ray B, Whitmore WF Jr (1975). Advanced prostatic carcinoma: flutamide therapy after conventional endocrine treatment. Urology 6:164-166.

Stoliar B, Albert DJ (1974). SCH 13521 in the treatment of advanced carcinoma of the prostate. J Urol 111:803-807.

Straffon RA, Kiser WF, Robitaille M, Dohn DF (1968). ^{90}Yttrium hypophysectomy in the management of metastatic carcinoam of the prostate gland in 13 patients. J Urol 99:102-105.

The Leuprolide Study Group (1984). Leuprolide versus diethylstilbestrol for metastatic prostate cancer. New Engl J Med 311:1281-1286.

Tisell LE, Salander H (1975). Androgenic properties and adrenal depressent activity of megestrol acetate observed in castrated male rats. Acta Endocrinol. 78:316-324.

Tomlinson RL, Currie DP, Boyce WH (1977). Radical prostatectomy: palliation for stage C carcinoma of the prostate. J Urol 117: 85-87.
Veterans Administration Cooperative Urological Research Group (VACURG) (1967). Treatment and survival of patients with cancer of the prostate. Surg Gynecol Obstet 124:1011-1017.
Warner B, Worgul TJ, Drago J, Demers L, Dufau M, Max D, Santen RJ, Abbott Study Group (1973). Effect of very high doses of D-Leucine^6-gonadotropin-releasing hormone pro-ethylamide on the hypothalamic-pituitary testicular axis in patients with prostatic cancer. J Clin Invest 72:1842-1855.
Waxman JH, Was JAH, Hendry WF, Whitfield HN, Besser GM, Malpas JS, Oliver RTD (1983). Treatment with gonadotropin-releasing hormone analogue in advanced prostatic cancer. Brit Med J 286:1309-1312.
Williams G, Blooms SR (1984). Treatment of advanced carcinoma of the prostate. Brit Med J 11:572-572.

AROMATASE AND AROMATASE INHIBITORS: FROM ENZYMOLOGY TO SELECTIVE CHEMOTHERAPY

M.E. Brandt, D. Puett, R. Garola, K. Fendl, D.F. Covey* and S.J. Zimniski.

The Reproductive Sciences and Endocrinology Laboratories, Dept. of Biochemistry, Univ. of Miami School of Medicine, Miami, FL 33101 and *Dept. of Pharmacology, Washington Univ. School of Medicine, St. Louis, MO 63110.

INTRODUCTION

Estrogens, most notably estradiol, play critical roles in the regulation of steroidogenesis and reproductive function and are responsible for the development of sexual differentiation in the brain (Williams et al., 1979; Brinkmann et al., 1980; Nozu et al., 1981; Naftolin and MacLusky, 1982; Ryan, 1982; Weisz, 1982; Fanjul et al., 1984; Roselli et al., 1984). As aromatic compounds with structures similar to a number of known carcinogens, estrogens have also been implicated in the induction and growth of a number of tumors, of which breast cancer is the most common (Huseby, 1980; Ryan, 1982; Santen, 1982). For these reasons, elucidation of the regulation of estrogen biosynthesis, and more specifically aromatase, has been of considerable interest.

Aromatase is found in highest concentration in the placenta, ovary, and testis, but unlike the majority of the steroidogenic enzymes, its localization is not limited to cells engaged in steroid biosynthesis. In addition to the above tissues, the enzyme is distributed throughout the body, most notably in adipose tissue, muscle, skin, and the central nervous system (Perel and Killinger, 1979; Longcope, 1982; Naftolin and MacLusky, 1982; Weisz, 1982; Berkovitz et al., 1984). In males and postmenopausal females, about 80% of estrogen synthesis occurs in peripheral tissues; even in the premenopausal female, 40% of aromatization occurs in extragonadal tissue (Kirschner et al., 1982).

CHARACTERISTICS OF THE AROMATASE ENZYME

Although it has been recognized for some time that estrogens are synthesized by the aromatization of androgens (Figure 1), the actual mechanism of this process has not been completely elucidated. Until the introduction of human placental microsomes as the enzyme source (Ryan, 1959) progress was relatively slow due to the fairly low activity in most tissues.

Ryan (1959) first established that the aromatization reaction required oxygen and NADPH; it was later shown that three molecules of both oxygen and NADPH were utilized (Thompson and Siiteri, 1974a). It has since been demonstrated that the aromatization process is catalyzed by one enzyme, aromatase, a cytochrome P-450 (Thompson and Siiteri, 1974b) which requires the presence of a cytochrome c reductase for activity. Members of this enzyme class usually catalyze hydroxylations, thus adding support to the initial observation by Meyer (1955), later confirmed by Ryan (1959), that 19-hydroxyandrostenedione is a putative intermediate in the process of aromatization. The 19-aldehyde form of androstenedione, produced from two sequential hydroxylations at the C-19 position, was later proposed to be the second intermediate (Akhtar and Skinner, 1968). Presently, there is considerable evidence that the first two oxidations in aromatization are hydroxylations. The nature of the final oxidation is still in question, but it has been proposed to be another hydroxylation (Goto and Fishman, 1977). The product of this step, 2β-hydroxy-19-oxo-androstenedione, was found to rapidly collapse non-enzymatically to estrone (Hosoda and Fishman, 1974). However, it has been shown that the fate of the various oxygens in the chemically synthesized 2β-hydroxy-19-oxo-androstenedione differs from the fate of the equivalent oxygens in the enzymatic aromatization of androstenedione (Caspi et al., 1984), thus raising considerable doubts about this hypothesis.

Aromatase, as a membrane-bound enzyme, becomes quite unstable upon separation from its normal milieu

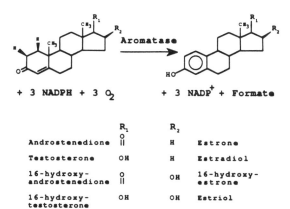

	R_1	R_2	
Androstenedione	=O	H	Estrone
Testosterone	OH	H	Estradiol
16-hydroxy-androstenedione	=O	OH	16-hydroxy-estrone
16-hydroxy-testosterone	OH	OH	Estriol

Figure 1. The reactions catalyzed by aromatase. The aromatization of the A ring of the steroid involves the loss of the C-19 methyl group and the 1β and 2β hydrogens. The reaction can be followed by measuring estrogen production, or, more commonly, by following production of tritiated water from substrate labeled with tritium in the 1β and 2β positions.

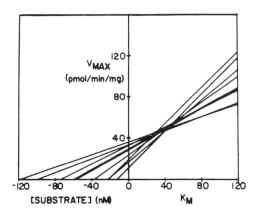

Figure 2. Direct linear plot for a typical kinetic experiment with placental microsomal aromatase. The lines are generated by plotting for each data point the concentration as the (-) abscissa intercept and the velocity as the ordinate intercept, and then connecting those points. The enzyme parameters are determined from the median value of the intersection of the lines (Eisenthal and Cornish-Bowden, 1974): The K_m from this plot is 51.6 nM, and the V_{max} is 66.0 pmol estrogen/min/mg microsomal protein.

and, therefore, has proven resistant to purification. However, at least five groups have obtained homogeneous or nearly homogeneous preparations (Osawa et al., 1982; Mendelson et al., 1985; Nakajin et al., 1986; Hagerman, 1987; Kellis and Vickery, 1987). The highly purified enzyme has a molecular weight of 52-55 kilodaltons.

Polyclonal antibodies raised against partially purified aromatase have been used to screen a λ gt11 expression library, and a partial cDNA clone coding for approximately 60% of the carboxy-terminal portion of aromatase has been obtained (Evans et al., 1986a). The amino acid sequence deduced from this clone suggests that aromatase is a member of a new cytochrome P-450 family, and this clone is currently being used as a probe to examine regulation of aromatase gene expression in various tissues (Evans et al., 1986b; Merrill et al., 1987).

Most of the kinetic data available for aromatase were obtained using a placental microsomal fraction. Differences in preparation methods, and possibly variations in the enzyme, have resulted in a range of values for the kinetic parameters.

Figure 2 shows the results of a typical kinetic experiment using human placental microsomes with testosterone as substrate. The data were analyzed by the direct linear plot (Eisenthal and Cornish-Bowden, 1974). The mean K_m ($\pm$ SEM) from several determinations was 43.6 $\pm$ 3.9 nM, while the calculated maximum velocity varied in different preparations between 30 and 70 pmol estrogen/min/mg microsomal protein.

The widely used tritiated water release assay for aromatase (Thompson and Siiteri, 1974a; Rabe et al., 1982; Zimniski et al., 1985) has the advantages of simplicity and speed over direct determination of the amount of estrogen formed; however, it requires confirmation that the rate of tritiated water production accurately reflects the rate of estrogen production. One of the methods used to address this issue has been the application of HPLC to separate the metabolites produced during incubation of testosterone with microsomal fractions. In placental microsomes, testosterone is normally converted to the estrogens,

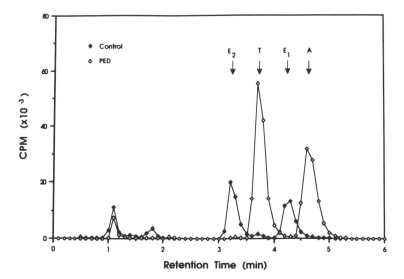

Figure 3. HPLC chromatogram of (1β,2β-^{3}H)testosterone metabolism by placental microsomes. The reaction was carried out in the absence (◆) or presence (◇) of 1.5 μM PED. The arrows denote the migration position of standards: E$_2$,estradiol; T,testosterone; E$_1$,estrone; and A,androstenedione. PED strongly inhibits the formation of estradiol and estrone. In the presence of PED, androstenedione is the major product; in contrast, nearly all of the available androgen is converted to estrogen in the absence of PED.

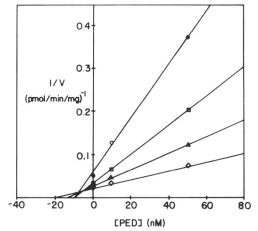

Figure 4. Dixon plot for placental aromatase. Each line represents a single testosterone concentration: 18.0 nM (○), 35.8 nM (□), 57.6 nM (△), and 90.5 nM (◇). The intersection of the lines is indicative of competitive inhibition.

estradiol and estrone, while in the presence of a potent enzyme inhibitor this substrate is primarily converted to androstenedione by the reversible enzyme 17β-hydroxysteroid dehydrogenase (Figure 3). In addition, there are a few minor metabolites migrating near the solvent front. Therefore, this assay is valid for placental microsomes, however, verification must be performed for each tissue.

AROMATASE INHIBITORS

Since estrogen production is totally dependent upon a functional aromatase, there is considerable interest in developing specific and efficacious enzyme inhibitors as a modality for the treatment of estrogen-dependent tumors (Brodie and Santen, 1987), as well as other estrogen-mediated disorders. Two agents, aminoglutethimide and Δ^1-testolactone, are currently in clinical use, but they are relatively non-specific and exhibit relatively low potencies as inhibitors of aromatase (Johnston and Metcalf, 1984).

The deficiencies of these compounds as specific and effective aromatase inhibitors have led to the development of improved enzyme inhibitors. One of these, 4-hydroxyandrostenedione (4-OH-A), initially reported by Brodie et al. (1976), is currently undergoing clinical trials in England (Coombes et al., 1984; Brodie et al., 1986; Goss et al., 1986). Brueggemeier and co-workers (Brueggemeier et al., 1982; Snider and Brueggemeier, 1987) have shown that 7α-derivatives of androstenedione are also highly effective aromatase inhibitors. Another inhibitor, 10-propargylestr-4-ene-3,17-dione (PED), was synthesized by Covey et al. (1981) and other laboratories (Metcalf et al., 1981; Marcotte and Robinson, 1982) and is currently being evaluated by Johnston and co-workers (cf. Johnston et al., 1984; Johnston, 1987) and by our laboratories (Zimniski et al., 1985, 1987; Puett et al., 1986).

in vitro Studies

We have extensively investigated the properties of PED in microsomal fractions from human placenta and rat ovaries (cf. Zimniski et al., 1987). In placental aromatase PED has proved to be an effective inhibitor

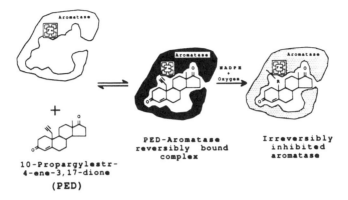

Figure 5. The action of a mechanism-based inactivator of aromatase. The inhibitor (e.g., PED) binds to the enzyme in a reversible fashion and then is thought to be activated by the enzyme in a rate-limiting step. The activated inhibitor then covalently binds to the enzyme, resulting in a permanently occupied active site.

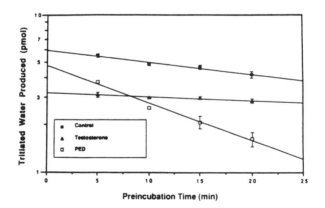

Figure 6. Spontaneous and induced inactivation of placental aromatase. The control points represent microsomes preincubated in the presence of NADPH, but in the absence of both substrate and inhibitor. The activity decreases with preincubation time; this inactivation is nearly abolished by the presence of 1.9 µM unlabeled testosterone which acts as a competitive inhibitor, hence the lower activity. Microsomes preincubated in the presence of 8.4 nM PED show a significantly higher rate of inactivation than do the control microsomes, indicating that PED is causing the loss of activity.

with 50% inhibition occurring at about 10 nM and greater than 95% inhibition being achieved at 150 nM. Dixon plots (Dixon, 1953) of velocity^{-1} versus PED concentration at varying concentrations of testosterone yielded straight lines intersecting at a common point in the -x,+y quadrant (Figure 4), a result indicative of competitive inhibition. The apparent competitive K_i from these data was calculated to be 4.1 ± 0.7 nM.

Both PED and 4-OH-A have been shown to be mechanism-based irreversible inactivators (suicide inhibitors) of human placental aromatase. These inhibitors require activation of a latent reactive moiety by the enzyme, following which they bond covalently to the active site of the enzyme, thereby inactivating it (Figure 5). This phenomenon can be addressed by experiments in which the microsomal preparation is preincubated with NADPH and various concentrations of PED for different periods of time and then analyzed for the amount of activity remaining. In the course of these experiments, we observed that the aromatase activity decreased with preincubation time even in the absence of PED. This decrease is presumably attributable to instability of the enzyme since the effect is prevented by the presence of the substrate during the preincubation (Figure 6). Also, the rate of decrease in activity in the presence of PED is significantly greater than the non-specific deterioration in activity in the absence of any added steroid.

The effects of PED on rat ovarian microsomal aromatase were also examined and compared with the human enzyme. In general, the results were similar to those observed with the human microsomal enzyme, although the rat enzyme has about a 3-fold lower affinity for PED than does the human enzyme. In addition, we have shown that PED is an irreversible inhibitor of the rat ovarian enzyme. Figure 7 shows that, in the presence of PED, the activity remaining decreases with increasing preincubation time in enzymes from both sources. The difference in y-intercepts in the lines in Figure 7 is due to the relative amount of enzyme bound to PED; the reversible binding to the enzyme is diffusion-limited, while inactivation is considerably slower. The time-dependent K_i, corrected for deterioration of the enzyme

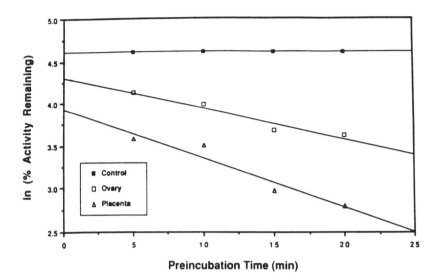

Figure 7. A semi-log plot from the time-dependent inhibition experiments on the human placental and PMSG-stimulated rat ovarian aromatase. Inactivation of the enzyme is a pseudo-first order process, dependent on the concentration of reversibly bound inhibitor-enzyme complex. The slope of the semi-log plot of percent active enzyme remaining versus preincubation time is $-k'$, where $k'=k_{inact} [I] /(K_i+ [I])$ (Kitz and Wilson, 1962). This type of plot is used to determine the k' for several inhibitor concentrations (83.6 nM PED is shown here). The rectangular hyperbola thus generated can then be analyzed for the rate constant of inactivation, k_{inact}, and the K_i by the direct linear plot. Both the human and rodent enzymes (after correction for spontaneous deterioration of activity) have significant rates of inactivation. The difference in ordinate intercepts is due to the higher affinity of the human enzyme for PED; the enzymes are inactivated with similar half-lives (12 vs. 16 min.).

in the absence of PED, was calculated to be 5 nM in the placental microsomes and 15 nM in the ovarian microsomes.

Since activation of the inhibitor by the enzyme is required for covalent bond formation, the inhibitors are expected to be specific for aromatase. However, both PED and 4-OH-A are androstenedione analogs and thus susceptible to metabolism by steroidogenic enzymes. Moreover, these steroid inhibitors, or their metabolites, are also potential ligands for steroid receptors. Previous research has demonstrated that PED and 4-OH-A display nominal hormonal activity in bioassays (Brodie et al., 1977, Johnston et al., 1984), and Johnston et al. (1984) further reported that PED binds only weakly to estrogen and androgen receptors.

Using a human ovarian carcinoma we have confirmed that PED does not compete for the androgen receptor (Figure 8), but 4-OH-A was found to compete slightly at 1 µM. These data are consistent with earlier reports which suggested that 4-OH-A is a weak androgen and binds to androgen receptors (Eil and Edelson, 1984; Hsiang et al., 1987). In our hands, neither PED nor 4-OH-A were found to compete with estradiol for the rat uterine estrogen receptor (Figure 9).

In contrast to these results, MacIndoe et al. (1982) reported that 4-OH-A does compete somewhat for cytosolic estrogen binding sites in human mammary tumors (MCF-7) cells. Furthermore, they found that 4-OH-A, when incubated with cultured mammary tumor cells, was equipotent to estradiol in inducing the progesterone receptor and in causing the depletion of cytosolic estrogen receptors. Wing et al. (1985) have also observed a significant uterotropic effect of 4-OH-A in ovariectomized rats following prolonged administration. This effect was attributed to the weak androgenic activity of 4-OH-A, a conclusion supported by the reversal of the uterotropic action by an antiandrogen (Wing et al., 1984). In comparison, PED showed some in vivo uterotropic activity (Johnston et al., 1984), but only at doses 3-fold higher than 4-OH-A. The observed discrepancies in apparent hormonal activity of the inhibitors may be due to the inherent differences among

Aromatase and Aromatase Inhibitors / 75

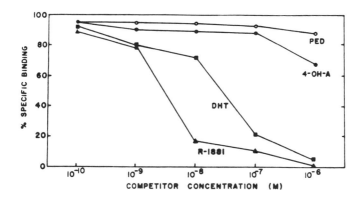

Figure 8. Competition of aromatase inhibitors with the human androgen receptor. Cytosol from a human ovarian carcinoma (BG-1) was assayed for specific binding of ^{3}H-R-1881 (methyltrienolone) following incubation of triplicate samples at 4 C for 2 hours with increasing concentrations of unlabeled R-1881 (▲), dihydrotestosterone (■), or the aromatase inhibitors 4-OH-A (●) and PED (○). The competition is reported as percent specific binding.

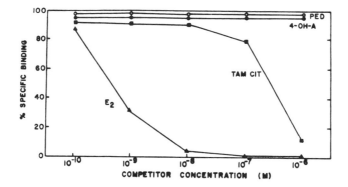

Figure 9. Competition of aromatase inhibitors with the rat estrogen receptor. Cytosol from rat uteri were incubated as described in Figure 8. In order to assess whether the aromatase inhibitors competed for binding of ^{3}H-estradiol to its receptor, increasing concentrations of unlabeled estradiol (▲), tamoxifen citrate (■), 4-OH-A (●), or PED (○) were added to parallel assays. All assays were performed in triplicate and results are reported as percent specific binding. No competition was observed with either inhibitor.

in vitro receptor assays, cell culture, and in vivo bioassays. Since the major differences have occurred in whole cells or in vivo, the effects may be mediated by metabolites of these inhibitors rather than the native compounds.

in vivo Studies

In non-pregnant, premenopausal women the ovary is the major source of the estrogens; thus ovarian aromatase is of considerable interest in the studies of hormone-dependent cancer and the role of estrogen biosynthesis in reproductive endocrinology. Also, in the rat the ovary is the most concentrated source of aromatase and therefore the most useful indicator of aromatase inhibitor effectiveness in vivo.

We have conducted experiments to examine the effects of PED in adult proestrous rats. Following incubation of ^{3}H-testosterone with rat ovarian microsomes from control animals, a number of metabolites (as yet uncharacterized) can be identified in addition to estradiol (Figure 10, upper panel). A single injection of PED, 12 hours prior to removal of the ovaries is sufficient to almost abolish estrogen production in vitro (Figure 10, lower panel). In addition, the formation of some of the metabolites is also sensitive to PED. These results demonstrate the effectiveness of PED as an aromatase inhibitor in the intact animal.

One method of treatment of estrogen-dependent breast tumors has involved the use of aromatase inhibitors. By virtue of their ability to decrease the amount of estrogen available to support the growth of tumor cells either directly or indirectly, specific inhibitors of aromatase have the potential to be effective with a minimum of side effects. Of the compounds clinically approved as anti-tumor drugs, two have as their apparent mechanism-of-action the inhibition of aromatase. Neither, however, is an especially effective inhibitor. The more potent of the two, aminoglutethimide, is also an inhibitor of the first step of steroidogenesis, cholesterol side chain cleavage, and therefore inhibits the production of all steroid hormones (Dexter et al., 1967; Salhanick, 1982).

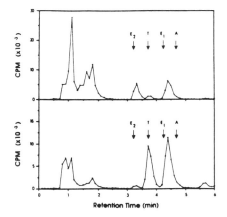

Figure 10. Chromatograms of testosterone metabolism by ovarian microsomes. The upper panel shows the metabolism by microsomes from control proestrus rats, while the lower panel shows data from proestrus rats sacrificed 12 hours after a single subcutaneous injection of 34 mg PED/kg body weight. PED inhibits production of estradiol in vivo as well as in vitro, it also inhibits the formation of a number of other, currently unidentified metabolites migrating near the solvent front.

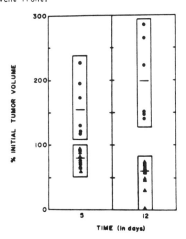

Figure 11. Regression of DMBA-induced rat mammary tumors in response to PED. Adult female rats bearing several mammary tumors were injected with either PED (5 mg/kg/day) or the sesame oil vehicle for two weeks. Tumors in control animals (●) continued to grow throughout the treatment period. In contrast, tumors in PED-treated animals (▲) regressed during treatment. After 12 days tumors in treated animals had decreased to 60% of their original size, while tumors in control animals had doubled in size.

Thus, hormone replacement therapy is required in clinical usage, although some unpleasant side effects have been noted. The other inhibitor, Δ^1-testolactone, appears to have about the same efficacy as aminoglutethimide (MacIndoe et al., 1982; Johnston and Metcalf, 1984).

Recently, the more specific aromatase inhibitors, 4-OH-A and PED, have been tested as potential anticancer drugs. Brodie et al. (1977) first demonstrated that 4-OH-A is effective against estrogen-dependent tumors in animals. That compound has now been used in clinical trials (cf. Brodie et al., 1986). Coombes et al. (1984) initially reported that 4-OH-A was a promising new drug for the treatment of advanced breast cancer. In phase II studies, they have further shown that 4-OH-A is effective in the treatment of postmenopausal breast cancer without any significant systemic side effects (Goss et al., 1986).

We have initiated studies to evaluate the efficacy of PED against endocrine-responsive tumors. The administration of PED to rats bearing the dimethylbenzanthracene-induced mammary tumor led to significant regression of the hormone-dependent tumors (Figure 11). These results are sufficient to suggest that PED may prove useful for clinical studies.

CONCLUSIONS

The use of aromatase inhibitors and antiestrogens in the treatment of advanced breast cancer represents a marked departure from surgery and non-specific chemotherapeutic agents (Jordan, 1985; Brodie and Santen, 1987). The specificity that can be achieved with these endocrine-related therapies ensures that side effects are minimal; moreover, combinational therapy involving, for example, an aromatase inhibitor and a "pure" antiestrogen should prove even more effective, and again with no known serious complications.

The clinical success enjoyed by aminogluthethimide and Δ^1-testolactone, two rather non-specific aromatase inhibitors, speaks well for the potential of the new generation of inhibitors, particularly the mechanism-based suicide inhibitors. As demonstrated by the work reported herein and in other laboratories, PED is an extremely potent aromatase inhibitor both in vitro and in vivo. In addition, it has extremely weak, if any, interaction with estrogen and androgen receptors. We have shown its effectiveness in inhibiting the growth of DMBA-induced mammary tumors in adult rats, and Zimniski and co-workers found that it also inhibits growth of a human ovarian carcinoma maintained in female nude mice (Garola et al., 1987).

In conclusion, PED, 4-OH-A, and other members of the new generation of aromatase inhibitors deserve a rapid and thorough evaluation of their effectiveness in treating human estrogen-dependent tumors.

ACKNOWLEDGMENTS

This work was supported by the National Institutes of Health (CA43226, CA23582, CA00829, DK33973, and RR05363), the Florida Division of the American Cancer Society Carl B. Ferguson Fund, the American Cancer Society Institutional (UM) Grant, and the Miami Women's Cancer Association.

REFERENCES

Akhtar M, Skinner SJM (1968). The intermediary role of 19-oxoandrogen in the biosynthesis of oestrogen. Biochem J **109**: 318-321.

Berkovitz GD, Fujimoto M, Brown TR, Brodie AM, Migeon CJ (1984). Aromatase activity in cultured human genital skin fibroblasts. J Clin Endocrinol Metab **59**: 665-671.

Brinkmann AO, Leemborg FG, Roodnaat EM, De Jong FH, van der Molen HJ (1980). A specific action of estradiol on enzymes involved in testicular steroidogenesis. Biol Reprod **23**: 801-809.

Brodie AMH, Santen RJ (1987). Aromatase in breast cancer and the role of aminoglutethimide and other aromatase inhibitors. CRC Crit Rev Oncology Hematology **5**: 361-396.

Brodie AMH, Schwarzel WC, Brodie HJ (1976). Studies on the mechanism of estrogen biosynthesis in the rat ovary-I. J Steroid Biochem **7**: 787-793.

Brodie AMH, Schwarzel WC, Shaikh AA, Brodie HJ (1977).
The effect of an aromatase inhibitor, 4-hydroxy-4-androstene-3,17-dione, on estrogen-dependent processes in reproduction and breast cancer. Endocrinology **100**: 1684-1694.

Brodie AMH, Wing LY, Goss P, Dowsett M. Coombes RC (1986). Aromatase inhibitors and their potential clinical significance. J Steroid Biochem **25**: 859-865.

Brueggemeier RW, Snider CE, Counsell RE (1982). Substituted C_{19} steroid analogs as inhibitors of aromatase. Cancer Res (suppl) **42**: 3334s-3337s.

Caspi E, Wicha J, Arunachalom T, Nelson P, Spiteller G (1984). Estrogen Biosynthesis. Concerning the obligatory intermediacy of 2β-hydroxy-10β-formylandrost-4-ene-3,17-dione. J Amer Chem Soc **106**: 7282-7283.

Coombes RC, Goss P, Dowsett M, Gazet JC, Brodie AMH (1984). 4-Hydroxyandrostenedione in treatment of postmenopausal patients with advanced breast cancer. Lancet **2**: 1237-1239.

Covey DF, Hood WF, Parikh VD (1981). 10β-Propynyl-substituted steroids: mechanism-based enzyme-activated irreversible inhibitors of estrogen biosynthesis. J Biol Chem **256**: 1076-1079.

Dexter RN, Fishman LM, Ney RL, Liddle GW (1967). Inhibition of adrenal corticosteroid synthesis by aminoglutethimide: studies of the mechanism of action. J Clin Endocrinol Metab **27**: 473-480.

Dixon M (1953). The determination of enzyme inhibitor constants. Biochem J **55**: 170-171.

Eil C, Edelson SK (1984). The use of human skin fibroblasts to obtain potency estimates of drug binding to androgen receptors. J. Clin Endocrinol Metab **59**: 51-55.

Eisenthal R, Cornish-Bowden A (1974). The direct linear plot: a new graphical procedure for estimating enzyme kinetic parameters. Biochem J **139**: 715-720.

Evans CT, Ledesma DB, Schulz TZ, Simpson ER, Mendelson CR (1986a). Isolation and characterization of a cytochrome P-450 mRNA. Proc Natl Acad Sci USA **83**: 6387-6391.

Evans CT, Steinkampf MR, Simpson ER, Mendelson CR (1986b). Regulation of estrogen biosynthesis is mediated by changes in the rate of synthesis of aromatase cytochrome P-450. Endocrine Soc 68th Ann Meeting 183 (Abstr. 609).

Fanjul LF, Galarreta CMR, Hsueh AJW (1984). Estrogen regulation of progestin biosynthetic enzymes in cultured rat granulosa cells. Biol Reprod **30**: 903-912.

Garola R, Brandt M, Fendl K, and Zimniski SJ (1987). Steroid receptors in a human ovarian carcinoma established in nude mice. Steroids, in press.

Goss PE, Powles TJ, Dowsett M, Hutchinson G, Brodie AMH, Gazet JC, Coombes RC (1986). Treatment of advanced breast cancer with an aromatase inhibitor, 4-hydroxyandrostenedione: phase II report. Cancer Res **46**: 4823-4826.

Goto J, Fishman J (1977). Participation of a nonenzymatic transformation in the biosynthesis of estrogens from androgens. Science **195**: 80-81.

Hagerman DD (1987). Human placenta estrogen synthetase (aromatase) purified by affinity chromatography. J Biol Chem **262**: 2398-2400.

Hosoda H, Fishman J (1974). Unusually facile aromatization of 2β-hydroxy-19-oxo-4-androstene-3,17-dione to estrone. Implications in estrogen biosynthesis. J Amer Chem Soc **96**: 7325-7329.

Hsiang Y-HH, Berkovitz GD, Brown TR, Migeon CJ, Brodie AMH (1987). The infuence of 4-hydroxy-4-androstene-3,17-dione on androgen metabolism and action in cultured human foreskin fibroblasts. J Steroid Biochem **26**: 131-135.

Huseby RA (1980). Demonstration of a direct carcinogenic effect of estradiol on Leydig cells of the mouse. Cancer Res **40**: 1006-1011.

Johnston JO (1987). Biological characterization of 10-(2-propynyl)estr-4-ene-3,17-dione (MDL 18,962), an enzyme-activated inhibitor of aromatase. Steroids, in press.

Johnston, JO, Metcalf BW (1984). Aromatase: a target enzyme in breast cancer. In: "Novel Approaches to Cancer Chemotherapy," New York: Academic Press, pp 307-328.

Johnston JO, Wright CL, Metcalf BW (1984). Biochemical and endocrine properties of a mechanism-based inhibitor of aromatase. Endocrinology **115**: 776-785.

Jordan CV (1985). Antiestrogens as antitumor agents. In: Hollander VP (ed): "Hormonally Responsive Tumors," New York: Academic Press, pp 219-235.

Kellis JT, Vickery LE (1987). Purification and characterization of human placental aromatase

cytochrome P-450. J Biol Chem **262**: 4413-4420.
Kirschner MA, Schneider G, Ertel NH, Worton E (1982).
Obesity, androgens, estrogens, and cancer risk.
Cancer Res (suppl) **42**: 3281s-3285s.
Kitz R, Wilson IB (1962). Esters of methanesulfonic
acid as irreversible inhibitors of
acetylcholinesterase. J Biol Chem **237**: 3245-3249.
Longcope C (1982). Methods and results of aromatization
studies in vivo. Cancer Res (suppl) **42**: 3307s-3311s.
MacIndoe JH, Woods GR, Etre LA, Covey DF (1982).
Comparative studies of aromatase inhibitors in
cultured human breast cancer cells. Cancer Res
(suppl) **42**: 3378s-3381s.
Marcotte PA, Robinson CH (1982). Design of mechanism-
based inactivators of human placental aromatase.
Cancer Res (suppl) **42**: 3322s-3326s.
Mendelson CR, Wright EE, Evans CT, Porter JC, Simpson ER
(1985). Preparation and characterization of
polyclonal and monoclonal antibodies against human
aromatase cytochrome P-450 (P-450$_{AROM}$), and their use
in its purification. Arch Biochem Biophys **243**: 480-491.
Merrill JC, Steinkampf MP, Mendelson CR, Simpson ER
(1987). FSH increases the levels of mRNA encoding
aromatase cytochrome P-450 in ovarian granulosa cells
of rat and human. Endocrine Soc 69th Ann Meeting 267
(Abstr. 984).
Metcalf BW, Wright CL, Burkhart JP, Johnston JO (1981).
Substrate-induced inactivation of aromatase by allenic
and acetylenic steroids. J Amer Chem Soc **103**: 3221-3222.
Meyer AS (1955). Conversion of 19-hydroxy-Δ^4-androstene-3,17-dione to estrone by endocrine tissue.
Biochim Biophys Acta **17**: 441-442.
Naftolin F, MacLusky NJ (1982). Aromatase in the
central nervous system. Cancer Res (suppl) **42**: 3274s-3276s.
Nakajin S, Shinoda M, Hall PF (1986). Purification to
homogeneity of aromatase from human placenta. Biochem
Biophys Res Commun **134**: 704-710.
Nozu K, Dufau ML, Catt KJ (1981). Estradiol receptor-
mediated regulation of steroidogenesis in
gonadotropin-desensitized Leydig cells. J Biol Chem
256: 1915-1922.
Osawa Y, Tochigi B, Higashiyama T, Yarborough C,
Nakamura T, Yamamoto T (1982). Multiple forms of

aromatase and response of breast cancer aromatase to antiplacental aromatase II antibodies. Cancer Res (suppl) **42**: 3299s-3305s.

Perel E, Killinger DW (1979). The interconversion and aromatization of androgens by human adipose tissue. J Steroid Biochem **10**: 623-627.

Puett D, Brandt ME, Covey DF, Zimniski SJ (1986). Characterization of a potent inhibitor of aromatase: Inhibition of the rat ovarian enzyme and regression of estrogen-dependent mammary tumors by 10-propargylestr-4-ene-3,17-dione. In Baulieu EE, Iacobelli S, McGuire WL (eds): "Endocrinology and Malignancy," Lancaster: Parthenon Publ., pp 279-289.

Rabe T, Rabe D, Runnebaum B (1982). New aromatase assay and its application for inhibitory studies of aminoglutethimide on microsomes of human term placenta. J Steroid Biochem **17**: 305-309.

Roselli CE, Ellinwood WE, Resko JA (1984). Regulation of brain aromatase activity in rats. Endocrinology **114**: 192-200.

Ryan KJ (1959). Biological aromatization of steroids. J. Biol Chem **234**: 268-272.

Ryan KJ (1982). Biochemistry of aromatase: significance to female reproductive physiology. Cancer Res (suppl) **42**: 3342s-3344s.

Salhanick HA (1982). Basic studies on aminoglutethimide. Cancer Res (suppl) **42**: 3315s-3321s.

Santen RJ (1982). Introduction to the conference, Aromatase: new perspectives for breast cancer. Cancer Res (suppl) **42**: 3268s.

Snider CE, Brueggemeier (1987). Potent enzyme-activated inhibition of aromatase by a 7α-substituted C_{19} steroid. J Biol Chem **262**: 8685-8689.

Thompson EA, Siiteri RK (1974a). Utilization of oxygen and reduced nicotinamide adenine dinucleotide phosphate by human placental microsomes during aromatization of androstenedione. J Biol Chem **249**: 5364-5372.

Thompson EA, Siiteri PK (1974b). The involvement of human placental microsomes during aromatization of androstenedione. J Biol Chem **249**: 5364-5372.

Weisz J (1982). In vitro assays of aromatase and their role in studies of estrogen formation in target tissues. Cancer Res (suppl) **42**: 3295s-3298s.

Williams MT, Roth, MS, Marsh JM, LeMaire WJ (1979).

Inhibition of human chorionic gonadotropin-induced progesterone synthesis by estradiol in human luteal cells. J Clin Endocrinol Metab **48**:437-440.

Wing L-Y, Tsai-Morris C, Brodie AMH (1984). The effect of aromatase inhibitor 4-hydroxyandrostenedione on sex steroid target tissue. Excerpta Medica **652**: 2535A.

Wing L-Y, Garrett WM, Brodie AMH (1985). Effects of aromatase inhibitors, aminoglutethimide, and 4 - hydroxyandrostenedione on cyclic rats and rats with 7,12-dimethylbenz(a)anthracene-induced mammary tumors. Cancer Res **45**:2425-2428.

Zimniski SJ, Brandt ME, Melner MH, Covey DF, Puett D (1985). Inhibition of Leydig tumor cell steroidogenesis by 10-propargylestr-4-ene-3,17-dione, an irreversible aromatase inhibitor. Cancer Res **45**: 4883-4889.

Zimniski SJ, Brandt ME, Covey DF, Puett D (1987). Inhibition of aromatase activity and of endocrine-responsive tumor growth by 10-propargylestr-4-ene-3,17-dione and its 17-propionate derivative. Steroids, in press.

ESTROGENS AND ANTIESTROGENS MEDIATE CONTRASTING TRANSITIONS IN ESTROGEN RECEPTOR CONFORMATION WHICH DETERMINE CHROMATIN ACCESS: A REVIEW AND SYNTHESIS OF RECENT OBSERVATIONS

Katherine Nelson, John R. van Nagell, Holly Gallion, Elvis S. Donaldson, and Edward J. Pavlik
Departments of Obstetrics and Gynecology, and Biochemistry
University of Kentucky College of Medicine, Lexington, Kentucky, 40536

BACKGROUND

After exposure to estradiol, a series of molecular events are initiated in the rodent uterus which program the physiological responses that culminate in tissue growth. Estrogen-induced proliferation also occurs in certain tumors, involving some of the same molecular components that have been identified in normal target tissues. In particular, proteins that specifically bind estrogen (referred to as "estrogen receptors"), acquire properties after binding ligand that result in stable interactions within chromatin (Jensen and DeSombre, 1972; O'Malley and Means, 1974; Katzenellenbogen and Gorski, 1975; Yamamoto and Alberts, 1976; Katzenellenbogen et al., 1979; Katzenellenbogen, 1980). Although recent evidence indicates that estrogen receptors are nuclear proteins (Sheridan et al., 1979; Martin and Sheridan, 1982; Welshons et al., 1984; King and Greene, 1984; McClellan et al., 1984; Gasc et al., 1984; Molinari et al., 1985; Welshons et al., 1985; Gravinis and Gurpide, 1986), little is known about the chromatin components with which estrogen receptors interact. In general these interactions are thought to involve specific DNA sequences, (Jost et al., 1984; Jost et al., 1985), the nuclear matrix (Barrack and Coffey, 1980; Simmen et al., 1984) and acidic nonhistone protein-DNA (Spelsberg et al., 1983) because estrogen receptors bind to these components in disrupted cell-free systems.

Antiestrogens are typically nonsteroidal compounds with a characteristic triphenylethylene structure that can inhibit or antagonize the action of estradiol (Katzenellenbogen et al., 1979; Clark and Peck, 1979). Because estrogenic and antiestrogenic ligands interact with estrogen receptors very similarly, clear perceptions of how antiestrogens actually achieve antagonism have been difficult to develop. For example, antagonists compete with ^{3}H-estradiol for estrogen receptors (Katzenellenbogen et al., 1979; Clark and Peck, 1979), activate estrogen receptors (DeBoer et al. 1981; Katzenellenbogen et al., 1981) and cause estrogen receptors to be retained in chromatin (Katzenellenbogen et al., 1979; Clark and Peck, 1979).

Despite the fact that the antiestrogen, tamoxifen, ("Nolvadex") has been widely used clinically for the treatment of estrogen-responsive tumors (Tagnon, 1977; Fabian et al., 1981; Manni, and Arafah, 1981), little is known about the mechanisms of antagonist action. Although the Z-isomer (trans-tamoxifen) has the ability to antagonize estradiol, and the C-isomer (cis-tamoxifen) has the properties of an agonist, both isomers are substrates for microsomal conversion to phenolic forms which have a higher affinity for estrogen receptors (Borgna and Rochefort, 1979,1981; Fromson et al., 1973). Only the Z-isomer becomes converted to a high affinity ligand (trans-4OH-tamoxifen) with relative binding equal to or greater than estradiol (Robertson et al., 1982; Borgna and Rochefort, 1979,1981; Fromson et al., 1973; Katzenellenbogen et al., 1984). Isomeric interconversions have been observed which explain functional reversals where cis-4OH-tamoxifen, an agonist, acquires the properties of an antagonist (Katzenellenbogen et al., 1984). Specific binding sites for antiestrogens exist ubiquitously (Kon, 1983; Sudo et al., 1983) and are subject to competition by antiestrogens but not by estrogens (Sudo et al., 1983). These specific antiestrogen binding sites can be discriminated from estrogen receptors by competition studies (Sudo et al., 1983) and by physical characterizations (Kon, 1983). Antiestrogen binding sites are localized predominantley in microsomes, and differ from estrogen receptors in terms of thermal stability, pH stability and protease sensitivity (Sudo et al., 1983). It is unlikely that antiestrogen specific binding sites are involved in directly mediating the growth moderating antagonism to estrogen because the affinities of various antiestrogens for these sites do not parallel their potencies as antiestrogens (Sudo et al., 1983; Miller and Katzenellenbogen, 1983) and because t-butylphenoxyethyl diethylamine demonstrated no functional antagonism despite its ability to occupy sites specific for antiestrogens, while showing no affinity for estrogen receptors (Sheen et al., 1985). Moreover, antiestrogen resistant variants of MCF-7 cells that remain responsive to estrogen appear similar to the parent line with respect to the expression of specific antiestrogen binding sites (Miller et al., 1984). As a consequence, in studies on mechanisms of antagonist action, it is important to be able to discriminate antiestrogen binding to estrogen receptors from binding to antiestrogen binding sites. This discrimination can be made using the high affinity antiestrogen, ^{3}H-4OH-tamoxifen. Estrogen receptors complexed with either high affinity agonist or antagonist appear similar with respect to chromatographic properties (Pavlik et al., 1985; Myatt and Whitliff,1986), sedimentation (Attardi and Happe, 1986), equilibrium and kinetic characteristics (Pavlik et al., 1986; Attardi and Happe, 1986) and binding to DNA-cellulose or ATP-Sepharose (Attardi and Happe, 1986; Rochefort and Borgna, 1981; Evans et al. 1982). In the presence of molybdate, receptors complexed with either estrogen or antiestrogen experience an inhibition of 4S to 5S transformation, as well as DNA binding; however, only receptors complexed with agonist demonstrate a slower rate of ligand dissociation after activation which can be inhibited by molybdate (Rochefort and Borgna, 1981). While receptors complexed with either agonist or antagonist have similar quantitative binding to DNA, the receptor affinity for DNA was two-fold lower when receptors were complexed with antagonists (Evans et al., 1982). Overall incapacitation of receptor activation by antagonists does not seem to occur (Nawatta et al., 1981). Approaches that can sensitively detect molecular differences have shown that receptors complexed with antagonists occurred as larger receptor forms (Pavlik et al.,1985; Eckert and Katzenellenbogen, 1982; Tate et al., 1984; Geier et al., 1985), exhibited different susceptibilities to proteolysis (Pavlik et al.,1985; Attardi and Happe, 1986), interacted differently with a polyclonal goat antibody (Tate et al., 1984), eluted differently during DEAE-Sephadex chromatography (Ruh et al., 1983) and demonstrated different aqueous two phase partitioning properties (Hansen and Gorski, 1986). These characterizations support the concept of ligand-mediated molecular transitions governing receptor conformation.

Clinical resistance to antiestrogens is a problem encountered in the treatment of breast cancer (Maas et al., 1980) that does not occur as the result of a shift from receptor-positive to receptor-negative status since only rarely are estrogen receptor-positive tumors associated with receptor negative metastases (Rosen et al., 1977). Moreover, cell-lines that have been selected for on the basis of antiestrogen resistance contain significant quantities of estrogen receptors which remain functionally able to mediate responses to agonists (Bronzert et al., 1985; Nawatta et al., 1981). Thus, a relevant aspect of resistance to antiestrogens is the occurrence of this failure in tissues which retain the expression of estrogen receptors.

Examples of resistance to antiestrogen may provide clues to the mechanisms of hormone action. Tamoxifen and 4OH-tamoxifen are as uterotrophic as estradiol in the mouse, but are reputed to have no antiuterotrophic activity (Jordan et al., 1978a,b; Terenius, 1970, 1971). Our own published experience (Pavlik et al., 1986) as well as that of a laboratory experienced in the use of the uterotrophic/antiuterotrophic assay indicate that several antiestrogens, including tamoxifen and 4OH-tamoxifen are indeed antiuterotrophic in the mouse (Hayes, 1981). These contrasting reports can be explained on the basis of dosage and the lower sensitivity of the mouse (Pavlik et al., 1986). Thus, because mice respond to lower doses of estradiol than rats, 4OH-tamoxifen is antiuterotrophic only when simultaneously injected with a sufficiently low concentration of agonist (Pavlik et al., 1986; Hayes, 1981). These phenomena suggest that under certain conditions, normal physiological systems can fail to respond to antagonists and can present well-defined models for examining resistance to antiestrogens.

Hormonal treatment also appears to remodel tissue responsiveness so that estrogen antagonism by antiestrogens is lost. Tamoxifen is a complete antagonist in the chick oviduct, inhibiting the estrogen-induced synthesis of ovalbumin and conalbumin (Sutherland et al., 1977). Administered alone, tamoxifen does not increase ovalbumin synthesis and only slightly increases conalbumin synthesis (Sutherland et al., 1977). However, under the influence of progestins or glucocorticoids, tamoxifen acquires the capacity for amplifying ovalbumin and conalbumin synthesis (Schweizer et al., 1986; Boue et al., 1985). Thus, the administration of progestins and glucocorticoids appears to reverse responses to tamoxifen and bring about a hormone induced resistance to antiestrogen.

PERSPECTIVES: CHROMATIN ACCESS LIMITED BY CONTRASTING TRANSITIONS IN ESTROGEN RECEPTOR CONFORMATION

Contrasting Receptor Transitions Mediated by Different Ligands

Physical characterizations of estrogen receptors complexed with estradiol and 4OH-tamoxifen, indicated that differences existed in the chromatographic behavior between receptors complexed with agonist or antagonist (Pavlik et al., 1985). These differences included the inability of cytoplasmic estrogen receptors complexed with ^{3}H-4OH-tamoxifen to be transformed to a smaller form in the presence of KCl or urea, as well as the appearance of larger forms of receptors complexed with 4OH-tamoxifen after exposure to trypsin. Extracted nuclear estrogen receptors complexed with 4OH-tamoxifen also chromatographed as hydrodynamically larger molecules than nuclear receptors complexed with estradiol. These observations have been interpreted to indicate that estradiol and 4OH-tamoxifen mediate contrasting transitions in estrogen receptor conformation. In this sense, the conformational transitions mediated by 4OH-tamoxifen result in intermolecular associations which become difficult to disrupt with KCl or urea and which alter the accessibility of trypsin to sensitive proteolytic sites. These observations are physiologically relevant because when receptor binding occurred within the intact uterus, extracted nuclear receptors complexed with antagonists were larger than receptors complexed with estradiol. These data indicate that mechanisms separating the action of estrogens and antiestrogens may result as a consequence of contrasting ligand-mediated receptor transitions. One result may be that estrogen receptors become positioned distinctly in chromatin by agonistic and antagonistic ligands. This positioning can be simply envisioned in terms of chromatin access limited by the larger size of estrogen receptors complexed with antagonists so that these receptor complexes become unable to reach certain chromatin regions. As a consequence, antagonism would result because the full range of events leading to an estrogenic response cannot be initiated (Pavlik et al., 1985).

Estrogen Receptor Complexed with ^{3}H-4OH-Tamoxifen Cannot Access Mg^{++}-Soluble Chromatin

DNAase I has been used for chromatin fractionation because these rapid enzymatic procedures disrupt only the contributions of DNA continuity to nuclear structure (Gottesfeld et al., 1974; Marushige and Bonner, 1971; Weintraub and Groudine, 1976; Garel and Axel, 1976; Senior and Frankel, 1978; Bloom and Anderson, 1978; Levy et al., 1979; Levy-Wilson and Dixon, 1979) and generate a Mg^{++}-soluble fraction to which template activity has generally been ascribed (Weintraub and Groudine, 1976; Garel and Axel, 1976). There is also considerable evidence that nuclease-mediated fractionation releases nuclear steroid receptors in association with nucleosome monomers, oligomers and linker DNA (Senior and Frankel, 1978; Rennie, 1979; Massol et al., 1978; Frankel and Senior, 1986), so that the potential exists for directly identifying chromatin components that interact with estrogen receptors. The value of nuclease-mediated chromatin fractionation has been shown to reside in an ability to generate a chromatin domain which accommodates estrogen receptors complexed with estradiol (Pavlik and Katzenellenbogen, 1982; Hemminki and Vauh Konen, 1977; Hemminki, 1977; Scott and Frankel, 1980), while restricting entry by estrogen receptors complexed with antiestrogens (Pavlik et al., 1986; Pavlik et al., 1985). There are a significant number of chromatin proteins in this fraction while only a fraction of nuclear estrogen receptors complexed with agonist are found in the Mg^{++}-soluble domain (Pavlik et al., 1986; Pavlik et al., 1985). This observation may be significant for several reasons. First, of the 15-20000 estrogen receptors/cell, 1-1500 remain in the nucleus six hours after hormonal exposure (Clark et al., 1973; Clark and Peck 1976), so that only a limited number of receptors may be involved in productive associations with chromatin, while the majority engage in nonproductive association (Yamamoto and Alberts, 1975; Gorski and Gannon, 1976). Second, receptors may have a dynamic passage through the Mg^{++}-soluble chromatin domain so that at any given time the opportunity for measurable receptor activity in this fraction is lower in contrast to fractions, like bulk chromatin, where receptors accumulate. Third, other presently unidentified chromatin regions may be found which individually also may appear to contain only a fraction of the total nuclear estrogen receptor activity, but may collectively account for a sizable share of the total nuclear receptor activity. Inherent in our research orientation is the concept that it is possible to define subpopulations of nuclear receptors which have significant roles in the mechanism of hormone action through the isolation of chromatin subfractions. An important aspect of this approach is the attempt to define relationships between estrogen receptors and chromatin components within intact tissue systems so that the opportunity for identifying physiologically significant processes may be maximized. In particular, the chromatin fractionation approach is well suited for examining mechanisms that separate the action of agonists and antagonists since the Mg^{++}-soluble chromatin fraction distinctively accommodates only estrogen receptors complexed with agonists.

Detailed observations on the reduced entry of estrogen receptors complexed with antagonist into Mg^{++}-soluble chromatin have been reported (Pavlik et al., 1986). These efforts show that tamoxifen and 4OH-tamoxifen are indeed capable of antagonizing estradiol in the mouse uterus. Poor penetration of the Mg^{++}-soluble chromatin domain was observed when intact uteri were exposed to antagonist. Prior exposure to bleomycin, an agent which induces strand breaks in DNA, did not reduce the activity of receptors complexed with agonists in this chromatin fraction indicating that a mechanism involving the interruption of DNA continuity by antagonist to restrict estrogen receptors from Mg^{++}-soluble chromatin is not likely. For delineating the Mg^{++}-soluble fraction which preferentially admits receptors complexed with agonists, DNAase I was superior to micrococcal nuclease indicating that each nuclease senses different features in chromatin. This superiority was supported by qualitative differences in the protein bands identified by SDS-PAGE. Increased ligand concentration could not be used to drive receptors complexed with antagonist into the Mg^{++}-soluble chromatin. Since more ^{3}H-4OH-tamoxifen penetrated free uterine cells than ^{3}H-estradiol, a restricted entry of antagonist does not contribute to the lack of penetration of the Mg^{++}-soluble chromatin. Tissue fractionation revealed that ^{3}H-4OH-tamoxifen activity was reduced in the high speed supernatant and increased in the microsomal/lysosomal pellet relative to ^{3}H-estradiol. However, these differences were not reflected in any decreased nuclear accumulation of ^{3}H-4OH-tamoxifen, so that extranuclear partitioning cannot be used to explain the dramatically reduced activity of receptors complexed with ^{3}H-4OH tamoxifen in Mg^{++}-soluble chromatin. However, this decrease does not occur through interference with receptor activation *per se* since receptors from these preparations demonstrated a significant capacity for binding to ATP-Sepharose. Even preparations which received molybdate during homogenization were characterized by lowered nuclear estrogen receptor indicating that molybdate influences the overall distribution of nuclear estrogen receptors (Pavlik et al., 1986).

Direct examination of *in situ* activation of nuclear estrogen receptors mediated by ^{3}H-4OH-tamoxifen or ^{3}H-estradiol within intact uteri has demonstrated that ATP-Sepharose binding was equivalent and extensive with both ligands

(Pavlik et al., 1987). Although some differences related to each ligand were noted in cytosol, these differences were minor and overshadowed by the more extensive receptor activation observed with both ligands in the extracted nuclear fraction. Consequently, the failure by estrogen receptors complexed with an antagonist to enter Mg^{++}-soluble chromatin cannot be interpreted as the result of gross differences that occur when agonists or antagonists mediate activation. Since extensive activation of nuclear receptors appeared similar after exposure to either agonist or antagonist and because tamoxifen, which ordinarily antagonizes the induction of ovalbumin/conalbumin synthesis by estrogen, acquires the ability to induce this synthesis after exposure to progesterone and glucocorticoids (Schweizer et al., 1986; Boue et al., 1985), it appears that receptors complexed with an antagonist may have complete functional competency. Thus, rather than viewing antagonism as resulting when receptors are rendered incompetent after interaction with an antiestrogen, it is appropriate to perceive the occurrence of functionally competent receptors complexed with antiestrogens which are restricted by ligand-mediated receptor transitions from chromatin regions where productive responses can be initiated. When this restriction is removed, as might occur after progestins and glucocorticoids are administered, and chromatin conformation is altered (Schweizer et al., 1986; Boue et al., 1985), these receptors are able to initiate estrogenic responses. There is considerable evidence that chromatin conformation is under hormonal control (Bloom and Anderson, 1983).

Recently, it has been reported that the ATP-mediated conversion of estrogen receptors to a smaller, lower affinity form in the presence of ^{3}H-estradiol, fails to occur in the presence of ^{3}H-4-OH-tamoxifen (McNaught et al., 1986). These observations also support the concept of distinctive ligand mediated transitions in estrogen receptor conformation and indicate that antiestrogens promote conformations that become insensitive to the effects mediated by ATP.

The concept that estrogen receptors complexed with antiestrogens are competent is strengthened enormously by observations that antiestrogens specifically stimulate the production of a 37 kDa secretory protein in MCF-7 cells (Sheen and Katzenellenbogen, 1987). The production of this protein is not stimulated by estrogens, which in fact inhibit its production. Since the production of other secretory proteins (32, 52, and 160 kDa) is stimulated by estrogen and inhibited by antiestrogens (Sheen and Katzenellenbogen, 1987), it is clear that estrogen receptors complexed with either agonist or antagonist are competent for the elicitation of responses involving secretory protein production.

It has been suggested that antiestrogens favor a high affinity receptor state which operates as a transcriptional repressor (McNaught et al., 1986). In light of the observation that antiestrogens are exclusively responsible for production of the 37 kDa secretory protein, it seems less likely that mechanisms involving receptor-mediated transcriptional repression are in operation. The fact that the secretory protein production elicited by estrogens and antiestrogens are quite different, indicates that receptors associated with antagonists may productively interact with chromatin sites that are distinct from those with which receptors associated with agonists are able to access. In this sense the chromatin access model not only must include the exclusion of receptors complexed with antiestrogens from regions where estrogenic responses are initiated, but must also incorporate the possibility that receptors complexed with antiestrogens access chromatin from which responses are initiated that result in products capable of exerting direct inhibitory control. An important inference from these observations is that receptors complexed with agonists cannot access chromatin from which inhibitory control is initiated. Nevertheless, these chromatin regions have not been demonstrated at present.

SUMMARY

Our perspective is that the mechanisms separating the action of estrogen agonists and antagonists involve contrasting transitions in receptor conformation which determine receptor access to different chromatin regions. Although we have suggested that size related exclusion of receptor may provide one manner for determining chromatin access (Pavlik et al., 1985), additional mechanisms may also be involved. The most recent evidence from several laboratories supports this perspective and indicates that estrogen receptors associated with antiestrogens are competent with respect to activation and response elicitation. Thus, we believe that an important course for future effort will be to define chromatin interactions that are limited to estrogen receptors complexed with either agonist or antagonist.

I. LITERATURE CITED:

Attardi B, Happe HK (1986). Comparison of the physicochemical properties of uterine nuclear estrogen receptors bound to estradiol or 4-hydroxytamoxifen. Endocrinology 119:904.
Barrack ER, Coffey DS (1980). Estrogen modulation of nuclear matrix-associated steroid hormone binding. J Biol Chem 255:7265.
Bloom KS, Anderson JN (1978). Fractionation of hen oviduct chromatin into transcriptionally active and inactive regions after selective micrococcal nuclease digestion. Cell 15:141-150.
Bloom KS, Anderson JN (1983). Hormonal regulation of the conformation of the ovalbumin gene in chick oviduct chromatin. J Biol Chem 257:13018.
Borgna JL, Rochefort H (1979). Occupation *in vivo* des recepteurs oestrogenes par des metabolites hydroxyles du tamoxifene. CR Acad

Sci (Paris) 289:1141.
Borgna JL, Rochefort H (1981). Hydroxylated metabolites of tamoxifen are formed *in vivo* and bound to estrogen receptor in target tissues. J Biol Chem 256:859.
Boue YL, Groyer A, Cadepond F, Groyer-Schweizer G, Robel P, Baulieu E-E (1985). Effects of progesterone and tamoxifen on glucocorticoid-induced egg-white protein synthesis in the chick oviduct. Endocrinology 116:2384.
Bronzert DA, Greene GL Lippman ME (1985). Selection and characterization of a breast cancer cell line resistant to the antiestrogen LY117018. Endocrinology 117:1409.
Clark JH, Anderson JN, Peck EJ (1973). Nuclear receptor estrogen complexes of rat uteri: concentration-time-response parameters. In O'Malley BW, Means AR (eds): "Hormone receptors and reproduction" vol 36 Adv Exp Biol Med, New York: Plenum Pub Co. pp 15-59.
Clark JH, Peck EJ, Jr (1976). Nuclear retention of receptor-oestrogen complex and nuclear acceptor sites. Nature 260:635.
Clark JH, Peck EJ, Jr (1979). "Triphenylethylene derivatives and estrogen antagonism". In Clark JH, Peck EJ (eds): "Monographs on Endocrinology: Female Sex Steroids," New York: Springer-Verlag, pp 99-134.
DeBoer W, Notides AC, Katzenellenbogen BS, Hayes JR, Katzenellenbogen JA (1981). The capacity of the antiestrogen CI-628 to activate the estrogen receptor *in vitro*. Endocrinology 108:206.
Eckert RL, Katzenellenbogen BS (1982). Physical properties of estrogen receptor complexes in MCF-7 human breast cancer cells. J Biol Chem 257:8840.
Evans E, Baskevitch PP, Rochefort H (1982). Estrogen-receptor-DNA interaction, difference between activation by estrogen and antiestrogen. Eur J Biochem 128:185.
Fabian C, Sternson L, El-Serafi M, Cain L, Hearne E (1981). Clinical pharmacology of tamoxifen in patients with breast cancer. Cancer 48:873.

Frankel FR, Senior MB (1986). The estrogen-receptor complex is bound at unusual chromatin regions. J Steroid Biochem 24:983

Fromsom JM, Pearson S, Bramah S (1973). The metabolism of tamoxifen (ICI 46,474) in laboratory animals. Xenobiotica 3:693-709 & 711.

Garel A, Axel R (1976). Selective digestion of transcriptionally active ovalbumin genes from oviduct nuclei. Proc Nat'l Acad Sci (USA) 73:3966

Gasc JM, Renoir JM, Radanyi C, Jaob I, Tuohimaa P, Baullieu EE (1984). Progesterone receptor in the chick oviduct: an immunohistochemical study with antibodies to distinct receptor components. J Cell Biol 99:1193.

Geier A, Haimsohn M, Beery R, Lunenfeld B (1985). Characterization of the 4-hydroxytamoxifen (4-OHTAM) bound estrogen receptor of MCF-7 cells solubilized by micrococcal nuclease. J Steroid Biochem 23:547.

Gorski J, Gannon F. (1976). Current models of steroid hormone action: a critique. Ann Rev Physiol 38:425.

Gottesfeld JM, Garrard WT, Bagi G, Wilson RF, Bonner J (1974). Partial purification of the template-active fraction of chromatin: a preliminary report. Proc Nat'l Acad Sci (USA) 71:2193.

Gravinis A, Gurpide E (1986). Enucleation of human endometrial cells: nucleo-cytoplasmic distribution of DNA polymerase α and estrogen receptor. J Steroid Biochem 24:469.

Hansen JC, Gorski J (1986). Conformational transitions of estrogen receptor monomer, effects of estrogens, antiestrogens, and temperature. J Biol Chem 261:13990.

Hayes, J.R. (1981) Bioactivities of antiestrogen and antiestrogen metabolites in the rat and mouse. Dissertation, University of Illinois, Champaign/Urbana, IL, p. 49.

Hemminki K (1977). Differential distribution of oestrogen receptors in subfractions of oviduct chromatin. Acta Endocrinol 84:215.

Hemminki K, Vauh Konen M (1977). Distribution

of estrogen receptors in hens oviduct chromatin fractions in the course of DNAase II digestion. Biochem Biophys Acta 474:109.
Jensen EV, DeSombre ER (1972). Mechanism of action of the female sex hormones. Ann Rev Biochem 41:203.
Jordan VC, Dix CJ, Naylor KE, Pristwich G, Rowsby L (1978). Non-steroidal antiestrogens: their biological effects and potential mechanisms of action. J Toxicology Environment Health 4:363.
Jordan VC, Rowsby L, Dix CJ, Prestwich G (1978). Dose related effects of non-steroidal antioestrogens and oestrogens on the measurement of cytoplasmic oestrogen receptors in the rat and mouse uterus. J Endocrinol 78:71.
Jost J-P, Geiser M, Seldraw M (1985). Specific modulation of the transcription of cloned avian vitellogenin II gene of estradiol-receptor complex *in vitro*. Proc Nat'l Acad Sci (USA) 82:988.
Jost J-P, Seldraw M, Geiser M (1984). Preferential binding of estrogen receptor complex to a region containing the estrogen-dependent hypomethylation site preceding the chicken vitellogenin II gene. Proc Nat'l Acad Sci (USA) 81:429.
Katzenellenbogen BS (1980). Dynamics of steroid hormone action. Ann Rev Physiol 42:17.
Katzenellenbogen BS, Bhakoo HS, Ferguson ER, Lan NC, Tatee T, Tsai TL, Katzenellenbogen JA (1979). Estrogen and antiestrogen action in reproductive tissues and tumors. Res Prog Hormone Res 35:259.
Katzenellenbogen BS, Gorski J (1975). Estrogen action on syntheses of macromolecules in target cells. In Litwack G (ed): "Biochemical Actions of Hormones" Volume 3, New York , Academic Press pp 187-243.
Katzenellenbogen BS, Norman MJ, Eckert RL, Peltz SW, Mangel WF (1984). Bioactivities, estrogen receptor interactions, and plasminogen activator-inducing activities of tamoxifen and hydroxy-tamoxifen isomers in MCF-7 human breast cancer cells. Cancer Res 44:112.

Katzenellenbogen BS, Pavlik EJ, Robertson DW, Katzenellenbogen JA (1981). Interaction of a high affinity anti-estrogen (α-[4-pyrrolidinoethoxy]phenyl-4-hydroxy-α'-nitrostilbene (CI 628M) with uterine estrogen receptors. J Biol Chem 256:2908.

King WW, Greene, GL (1984). Monoclonal antibodies localize oestrogen receptors in the nuclei of target cells. Nature 307:745.

Kon OL (1983). An antiestrogen-binding protein in human tissues. J Biol Chem 258:3173.

Levy WB, Connor W, Dixon GH (1979). Limited action of micrococcal nuclease on trout testis nuclei generates two mononucleosome subsets enriched in transcribed DNA sequences. J Biol Chem 254:609.

Levy-Wilson, B Dixon GH (1979). A subset of trout testis nucleosomes enriched in transcribed DNA sequences contains high mobility group proteins as major structured components. Proc Nat'l Acad Sci (USA) 76:1682.

Maass H, Jonat W, Stolzenback G, Trams G (1980). The problem of nonresponding estrogen receptor-positive patients with advanced breast cancer. Cancer 46:2835.

Manni A, Arafah BM (1981). Tamoxifen-induced remission in breast cancer by escalating the dose to 40 mg daily after progression on 20 mg daily. Cancer 48:873.

Martin PM, Sheridan PJ (1982). Towards a new model for the mechanism of steroid hormone action. J Steroid Biochem 16:215.

Marushige K, Bonner J (1971). Fractionation of liver chromatin. Proc Nat'l Acad Sci (USA) 68:2941.

Massol N, Lebeau MC, Baulieu EE (1978). Estrogen receptor in hen oviduct chromatin digested by micrococcal nuclease. Nucleic Acids Res. 5:723.

McClellan MC, West NB, Tachon DE, Greene GL, Brennen RM (1984). Immunocytochemical localization of estrogen receptors in the Maqcaque reproductive tract with antiestrophilins. Endocrinology 114:2002.

McNaught RW, Raymoure WJ, Smith RG (1986). Receptor interconversion model of hormone action 1. ATP-mediated conversion of estrogen receptors from a high to lower affinity state and its relationship to antiestrogen action. J Biol Chem 261: 17011-17017.

Miller MA, Katzenellenbogen BS (1983). Characterization and quantitation of antiestrogen binding sites in estrogen receptor-positive and negative human breast cancer cell lines. Cancer Res. 43:3094.

Miller MA, Lippman ME, Katzenellenbogen BS (1984). Antiestrogen binding in antiestrogen growth resistant estrogen-responsive clonal variants of MCF-7 human breast cancer cells. Cancer Research 44:5038.

Molinari AM, Medici N, Armetta I, Nigro V, Monchormont B, Puca GA (1985). Particulate nature of the unoccupied uterine estrogen receptor. Biochem Biophys Res Comm 128:634.

Myatt L, Wittliff JL (1986). Characterization of activated and non-activated estrogen and antiestrogen-receptor complexes by high performance ion exchange chromatography. J Steroid Biochem 24:1041.

Nawatta H, Bronzert D, Lippman ME (1981). Isolation and characterization of a tamoxifen-resistant cell line derived from MCF-7 human breast cancer cells. J Biol Chem 256:5016.

O'Malley BW, Means AR (1974). Female steroid hormones and target cell nuclei. Science 183:610.

Pavlik EJ, Katzenellenbogen BS (1982). The intranuclear distribution of rat uterine estrogen receptors determined after nuclease treatment and chromatin fractionation. Mol Cell Endocrinol 26:201.

Pavlik EJ, Nelson K, van Nagell JR Jr, Donaldson ES, Walden ML, Gallion H, Kenady DE (1987). Extensive *in situ* activation of nuclear estrogen receptors after exposure of murine uteri to [^{3}H] estradiol or [^{3}H]4-hydroxytamoxifen. Endocrinol 120:1608.

Pavlik EJ, Nelson K, van Nagell JR, Donaldson ES, Walden ML, Hanson MB, Gallion H, Flanigan RC, Kenady DE (1985). Hydrodynamic characterizations of estrogen receptors complexed with [^{3}H]-4-hydroxytamoxifen:

evidence in support of contrasting receptor transitions mediated by different ligands. Biochemistry 24:8101.

Pavlik EJ, van Nagell JR, Nelson K, Gallion H, Donaldson ES, Kenady DE, Baranowska-Kortylewicz J (1986). Antagonism to estradiol in the mouse: reduced entry of receptors complexed with 4-hydroxytamoxifen into a Mg^{++}-soluble chromatin fraction. Endocrinology 118:1924.

Rennie PS (1979). Binding of androgen receptor to prostatic chromatin requires intact linker DNA. J Biol Chem 254:3947.

Robertson DW, Katzenellenbogen JA, Long DJ, Rorke EA, Katzenellenbogen BS (1982). Tamoxifen antiestrogens. A comparison of the activity, pharmacokinetics, and metabolic activation of the *cis* and *trans* isomers of tamoxifen. J Steroid Biochem 16:1.

Rochefort A, Borgna JL (1981). Differences between oestrogen receptor activation by oestrogen and antioestrogen. Nature 292:256.

Rosen PO, Menendez-Botet CJ, Urban JA, Fracchia A, Schwartz MK (1977). Estrogen receptor protein (ERP) in multiple tumor specimens from individual patients with breast cancer. Cancer 39:2194

Ruh MF, Brzyski RG, Strange L, Ruh TS (1983). Estrogen and antiestrogen binding to different forms of the molybdate-stabilized estrogen receptor. Endocrinology 112:2203.

Schweizer G, Cadepond-Vincent F, Baulieu E-E (1986). Nuclear synthesis of egg white protein messenger ribonucleic acids in chick oviduct: effects of the anti-estrogen tamoxifen on estrogen-,progesterone-, and dexamethasone-induced synthesis. Biochemistry 24:1742.

Scott RW, Frankel JFR (1980). Enrichment of estradiol-receptor complexes in a transcriptionally active fraction of chromatin from MCF-7 cells. Proc Nat'l Acad Sci (USA) 77:1291.

Senior MB, Frankel FR (1978). Fractionation of hen oviduct chromatin into transcriptionally active and inactive regions after selective

micrococcal nuclease digestion. Cell 14:857.
Sheen YY, Katzenellenbogen BS (1987). Antiestrogen stimulation of the production of a 37,000 molecular weight secreted protein and estrogen stimulation of the production of a 32,000 molecular weight secreted protein in MCF-7 human breast cancer cells. Endocrinol 120:1140-1151.
Sheen YY, Simpson DM, Katzenellenbogen BS (1985). An evolution of the role oa antiestrogen-binding sites in mediating the growth modulatory effects of antiestrogens: studies using t-butyphenoxyethyl diethylamine, a compound lacking affinity for the estrogen receptor. Endocrinology 117:561.
Sheridan PJ, Buchanan JM, Anselmo VC (1979). Equilibrium: the intracellular distribution of steroid receptors. Nature 282:579.
Simmen RCM, Means AR, Clark JH (1984). Estrogen modulation of nuclear matrix-associated steroid hormone binding. Endocrinology 115:1197.
Spelsberg TC, Littlefield BS, Selke R, Dani GM, Toyoda H, Boyd-Leinen P, Thrall C, Kon OL (1983). Role of specific chromosomal proteins and DNA sequences in the nuclear binding sites for steroid receptors. Rec Prog Horm Res 39:463.
Sudo K, Monsoma FJ, Katzenellenbogen BS (1983). Antiestrogen binding sites distinct from the estrogen receptor: subcellular localization, ligand specificity and distribution in tissue in the rat. Endocrinol 112:425.
Sutherland R, Mester J, Baulieu EE (1977). Tamoxifen is a potent, pure anti-oestrogen in chick oviduct. Nature 267:434.
Tagnon HJ (1977). Antiestrogens in treatment of breast cancer. Cancer 39:2959.
Tate AC, Greene GL, Desombre ER, Jensen EV, Jordon VL (1984). Differences between estrogen- and anti-estrogen receptor complexes from human breast tumors identified with an antibody raised against the estrogen receptor. Cancer Res 44:1012.
Terenius L (1970). Two modes of interaction between oestrogen and anti-oestrogen. ACTA

Endocrinol 64:47.

Terenius L (1971). Structure activity relationships of anti-oestrogens with regard to interaction with 17B estradiol in the mouse uterus and vagina. ACTA Endocrinol 66:431.

Weintraub H, Groudine M (1976). Chromosomal subunits in active genes have an altered conformation. Science 193:848-856.

Welshons W, Krummel BM, Gorski J (1985). Nuclear localization of unoccupied receptors for glucocorticoids, estrogens and progestins in GH3 cells. Endocrinology 117:2140.

Welshons W, Lieberman ME, Gorski J (1984). Nuclear localizations of unoccupied oestrogen receptors. Nature 307:747.

Yamamoto KR, Alberts BW (1975). The interaction of estradiol-receptor protein with the genome: an argument for the existence of undetected specific sites. Cell 4:301.

Yamamoto KR, Alberts BW (1976). Steroid Receptors: elements for modulation of eucaryotic transcription. Ann Rev Biochem 45:721.

LONG-TERM TAMOXIFEN THERAPY TO CONTROL OR TO PREVENT
BREAST CANCER: LABORATORY CONCEPT TO CLINICAL TRIALS

V. Craig Jordan

Departments of Human Oncology and Pharmacology,
University of Wisconsin Clinical Cancer Center,
Madison, Wisconsin 53792

INTRODUCTION

Tamoxifen is a non-steroidal antiestrogen that has become the first line endocrine therapy for the palliative treatment of advanced breast cancer (Furr and Jordan, 1984). The development of tamoxifen is an intriguing story of serendipity and is an excellent example of how a drug targeted for one area can fail, and become extremely successful in another. Tamoxifen, originally known as ICI 46,474, was discovered as part of the fertility control program at Imperial Chemical Industries (ICI), Macclesfield, Cheshire, England. The drug was shown to be an excellent postcoital contraceptive in the rat (Harper and Walpole, 1966, 1967a,b). This excited interest in developing a "morning after" pill but clinical evaluation demonstrated that the concept could not be applied to humans. In fact, and somewhat ironically, tamoxifen induces ovulation in subfertile women and it is available in some countries as a profertility drug (Furr and Jordan, 1984).

The credit for the development of tamoxifen for the treatment of advanced breast cancer must go to the late Dr. Arthur Walpole who was the head of ICI's fertility control program. Walpole's group demonstrated the potent antiestrogenic activity of tamoxifen (Harper and Walpole, 1967a) and its ability to compete with estradiol for binding to the rat uterine estrogen receptor (Skidmore et al., 1972). The demonstration of estrogen receptors in human breast cancers and the identification of a correlation between estrogen receptor levels and the tumor response to endocrine therapy made non-steroidal antiestrogens an ideal therapy for evaluation.

Walpole had multiple reasons to suggest the use of tamoxifen to treat breast cancer. His early career had been dedicated to cancer research with a particular interest in carcinogenesis. In the mid 1940's Alexander Haddow in England (Haddow et al., 1944) showed that high-dose estrogen therapy could produce benefit for some women with advanced breast cancer. Walpole confirmed this preliminary report in a study with Dr. Edith Paterson at the Christie Hospital and Radium Institute in Manchester England (Walpole and Paterson, 1949). Various triphenylethylenes, made at ICI, were compared with diethylstilbestrol, a potent synthetic estrogen first described by Sir Charles Dodds a decade earlier (Dodds et al., 1938a,b). Dose-response effects were observed in the study but the reason for the apparently abitrary responses was elusive. Twenty years later, Walpole encouraged Mary Cole and her colleagues at the Christie Hospital to conduct a preliminary study to compare tamoxifen with diethylstilbestrol in the treatment of late advanced breast cancer (Cole et al., 1971). Subsequent studies around the world demonstrated that tamoxifen has a low incidence of side effects and proven efficacy. It is now the endocrine treatment of choice for advanced breast cancer (Furr and Jordan, 1984).

It is not the intention of this chapter to review all of the literature on antiestrogens. This information has been presented elsewhere (Sutherland and Jordan, 1981; Sutherland and Murphy, 1982; Jordan, 1984; Jordan, 1986a). Rather, the developing story of the current and future application of tamoxifen will be surveyed. The ideas for the uses of the drug have evolved as the strategic approaches to breast cancer therapy have changed over the past 15 years.

In the 1970's the strategy for breast cancer therapy changed from the application of chemotherapy to control advanced disease, to the use of adjuvant therapy after mastectomy. The aim was to cure patients by destroying micrometastases. Combination chemotherapy, though effective as a tumoricidal treatment, lacks tumor specificity and produces a spectrum of cytotoxic effects in normal tissues. The low incidence of side effects reported for tamoxifen during its testing as a therapy for advanced breast cancer made this a drug of choice for further evaluation. Nearly two decades ago, laboratory studies were started to determine the mode of action of tamoxifen as an antitumor agent in animal models. These data were necessary to plan treatment strategies and to formulate the best clinical application for the drug. This chapter will survey the evidence in favor of long-term adjuvant therapy with tamoxifen. Since the approach is proving to be successful, with very little patient morbidity, there is some interest in extending the use of tamoxifen to prevent the primary tumor from appearing. The potential use of tamoxifen as a possible preventative agent for breast cancer in women at high risk for the disease will be discussed, to formulate a logical strategy for the 1990's.

PHARMACOLOGICAL BASIS FOR LONG-TERM TAMOXIFEN THERAPY

The preliminary clinical report of the use of tamoxifen (then known as ICI 46,474) in late advanced breast cancer (Cole et al., 1971) acted as a catalyst to establish laboratory models to study the mode of action of the drug as an antitumor agent. The first studies (1973) were conducted at the Worcester Foundation for Experimental Biology using the dimethylbenzanthracene (DMBA) induced rat mammary carcinoma model (Jordan, 1974). The experiments were planned in collaboration with Dr. Arthur Walpole at ICI in England. The laboratory results demonstrated that tamoxifen inhibits the initiation and growth of DMBA-induced rat mammary carcinomata. These results and others (Jordan, 1975; Nicholson and Golder, 1975; Jordan, 1976a,b; Jordan and Dowse, 1976; Jordan and Koerner, 1976; Jordan and Jaspan, 1976; Nicholson et al., 1977a,b) were used to support the approval of tamoxifen for the treatment of advanced breast cancer in several countries. However, the clinical interest in adjuvant chemotherapy following mastectomy, stimulated a thorough examination of the mode of action of tamoxifen in the laboratory. Work conducted by Lippman (Lippman and Bolan, 1975) with the human breast cancer cell line MCF-7 in vitro, indicated that high concentrations of tamoxifen were tumoricidal. Therefore, it was believed that one or two years of adjuvant tamoxifen could cure some patients. The first model to demonstrate the potential of tamoxifen as a successful adjuvant therapy (Jordan, 1978) used animals (50 days old) which were given DMBA to produce microfoci of transformed mammary cells and then one month later, different daily doses of tamoxifen for 1 month (short course of therapy). The aim was to cure the animals with a short course of large doses of tamoxifen. These groups were compared with continuous tamoxifen therapy (long-term therapy). The results, first reported at a breast cancer symposium at Kings College, Cambridge in September 1977, demonstrated that only long-term tamoxifen therapy can maintain the animals tumor-free. Stopping tamoxifen therapy (short course) results in the appearance of tumors (Jordan et al., 1979). Dose-response studies with tamoxifen (short course) demonstrated that even very large (800 µg/daily) doses of the drug when given for a month (possibly equivalent to one year of therapy for patients) could not "cure" animals. Eventually the animals developed tumors once the drug was cleared from the body (Jordan and Allen, 1980; Jordan et al., 1980). Tamoxifen appears to have tumoristatic properties and is therefore effective only when drug administration is maintained.

The results in the DMBA-induced rat mammary carcinoma model have been confirmed and extended in the N-nitrosomethylurea (NMU)-induced mammary carcinoma model. Again, continuous therapy with tamoxifen prevents the appearance of tumors and short term therapy with large daily doses of tamoxifen does not cure the animals. Once therapy stops, or a short course of tamoxifen is given, the animals start to develop palpable tumors (Wilson et al., 1982; Jordan et al., 1984).

The evidence that tamoxifen produces a tumoristatic action in vivo is supported by the finding in vitro that tamoxifen can stop MCF-7 breast cancer cells from progressing through the cell cycle (Osborne et al., 1983; Sutherland et al., 1983). The block is at the G1/S interface and can be reversed with estrogen. High concentrations of tamoxifen do, however, produce effects which cannot be reversed with estrogen (Sutherland et al., 1986).

Current laboratory studies have focused upon the propagation of breast cancer cell lines in athymic mice (Soule and McGrath, 1980; Shafie and Grantham, 1980). The aim is to understand the mode of action of antiestrogens in vivo and also provide a model system to study the development of tamoxifen resistance. It is also possible that antiestrogen-dependent tumors may occur. Clearly this will provide an opportunity to design effective treatment strategies to use in the clinic when failure of tamoxifen therapy occurs.

However, the experimental approach is not entirely straight forward as tamoxifen has a complex, and often paradoxical, pharmacology (Jordan, 1984). Unlike the rat and human, tamoxifen is known to be an estrogen in the mouse (Furr and Jordan, 1984). Initially, concerns were expressed about the possibility that tamoxifen might be metabolized to estrogens (Jordan, 1982). This, however, does not appear to be the case (Lyman and Jordan, 1985; Jordan and Robinson, 1987). Furthermore, tamoxifen appears to exhibit a target site (not necessarily species) specificity so that, for example, the athymic mouse uterus may be stimulated to grow although there is no growth of implanted estrogen-dependent tumors (Jordan and Robinson, 1987).

There are only a few reports that describe the control of estradiol-stimulated tumor growth with tamoxifen in athymic mice (Osborne et al., 1985; Gottardis et al., 1985; Jordan et al., 1987). Tamoxifen produces a dose-related inhibition of estradiol-stimulated growth and tamoxifen cannot support the continued growth of established tumors after it is substituted for estradiol. However, tamoxifen does not appear to exhibit the properties of a tumoricidal agent. Long-term tamoxifen therapy appears to maintain the tumor cells in a state of suspended animation so that if tamoxifen therapy is stopped and estrogen treatment initiated, tumors regrow. These findings strongly support the position that tamoxifen therapy should be continued until therapeutic failure.

CLINICAL EVALUATION OF ADJUVANT TAMOXIFEN THERAPY

In Britain, the Nolvadex Adjuvant Trial Organization (NATO) entered its first patients on study in December 1977. The protocol was to treat with tamoxifen or not for two years and compare the recurrence rates of the disease (Baum et al., 1985). An attempt to evaluate the impact of extending the therapy with tamoxifen to five years (in the light of the already available animal data), was rejected in favor of the sound principle that an analysis of two years of tamoxifen would indicate whether there was to be any benefit at all. The results would also indicate whether a tumoricidal action could be demonstrated with short term therapy. Analysis of the clinical trial has proved to be extremely interesting, if not at times controversial. The NATO results indicate a disease-free survival advantage and an overall survival advantage for those patients receiving tamoxifen. Apparently there is no advantage for estrogen receptor positive vs estrogen receptor negative subgroups. This contrasts with virtually all other trials (Table 1).

In the United States the National Surgical Adjuvant Breast Project (NSABP) has compared combination chemotherapy (L-phenylalanine mustard plus 5-fluorouracil) with and without tamoxifen for two years in pre- and postmenopausal women. Early analysis showed an advantage for postmenopausal women (not premenopausal women) in the arm taking tamoxifen compared to chemotherapy alone (Fisher et al., 1981). There is a strong association between the response to tamoxifen and the estrogen and progesterone receptor status of the patient: those with high values of steroid receptors in the primary tumor have a longer disease-free interval than those with low values (Fisher et al., 1983).

The current trend towards meta-analysis, or overviews of all available clinical trials data, has revealed some important observations. As could be predicted from the laboratory studies, one year of adjuvant tamoxifen therapy is inferior to two years of adjuvant tamoxifen therapy (Richard Peto, personal communication). Furthermore, hazard analysis of individual trials like the NATO study show that the therapeutic effect of tamoxifen occurs only when the drug is being given (Professor Michael Baum, personal communication).

The first pilot study of long-term adjuvant tamoxifen therapy in stage II breast cancer patients was initiated in 1977 by Dr. Douglass C. Tormey at the University of Wisconsin Clinical Cancer Center. Up to that time, the available toxicology had been gathered from advanced breast cancer patients receiving one or perhaps two years of the drug. There were three aims of the study (Tormey and Jordan, 1984). Firstly, we wished to identify any unforeseen toxicities that had not previously been reported. Secondly, we wanted to determine the serum levels of tamoxifen and its metabolites during five years of treatment. Finally, and most importantly, we wanted to see whether there was any clinical benefit for patients on long-term tamoxifen therapy. We considered that five years of tamoxifen administration would be necessary, in

Table 1. Adjuvant Trials with Tamoxifen in Postmenopausal Patients.

Group	Daily Dose mg	Duration (yrs)	Increase in a) disease free interval	b) survival
Baum et al. (1985)	20	2	Yes	Yes
Ludwig (1984)	20	1	Yes	No
Pritchard et al. (1984)	30	2	Yes	No
Rose et al. (1985)	30	1	Yes	No
Wallgren et al. (1984)	40	2	Yes	No
Ribero and Palmer (1983)	20	1	Yes	No
Cummings et al. (1985)	20	2	Yes	No
Delozier et al. (1986)	40	3	Yes	Yes
Stewart & Prescott (1985)	20	5	Yes	Not Yet

the first instance, to determine the toxicology and
pharmacology of the drug before testing our concept in
large randomized clinical trials. Our pilot study compared three treatment arms that were similar in their
patient characteristics, but the patients were not
randomized to the arms. Combination chemotherapy was
compared to chemotherapy plus tamoxifen or chemotherapy
and tamoxifen with the tamoxifen continued for a further
four years (Tormey and Jordan, 1984). Tamoxifen was well
tolerated by the patients and long-term therapy did not
reveal any untoward side-effect that could be ascribed to
the drug. A high performance liquid chromatography assay
(Brown et al., 1983) for tamoxifen and its metabolites N-
desmethyl-tamoxifen and Metabolite Y (Jordan et al., 1983)
was established. Tamoxifen and its metabolites could be
measured in patients during long-term therapy and in
general, the serum levels remained stable throughout the
treatment period.

Although the study was only designed as a pilot, the
patients receiving long-term tamoxifen therapy have a low
recurrence rate for their disease. The decision was
therefore made to treat patients with tamoxifen indefinitely. The analysis after eight years of observation
demonstrates the benefit attained by the continuous
tamoxifen group (Tormey et al., 1987). There is a 71%
disease-free survival projected for the continuous
tamoxifen group, whereas 50% of patients receiving
chemotherapy had disease recurrence within two to three
years. When tamoxifen was given only when chemotherapy
was administered, the 50% recurrence rate was delayed a
further two years compared with chemotherapy alone.

A careful analysis of the endocrine consequences of
both chemotherapy and the combination with tamoxifen has
been made in this unique group of patients (Jordan et al.,
1987). In general, chemotherapy causes ovarian failure in
the majority of premenopausal patients but younger women
are more resistant (Dnistrian et al., 1983). As a result
young patients who continue tamoxifen may have an increased steroid secretion because of the direct ovarian
effects of tamoxifen (Groom and Griffiths, 1976; Manni and
Pearson, 1980; Sherman et al., 1979). This may not be an
optimal therapeutic situation because the estrogen may
reverse the action of tamoxifen. Indeed, pre-menopausal
patients with advanced breast cancer who are given tamoxifen and initially respond but ultimately failure, may

again respond to an endocrine manouver such as ovarian ablation (Sawka et al., 1986). Therefore, it seems logical that a competitive inhibitor of estrogen action like tamoxifen will function optimally in a low estrogen environment.

Overall, the adjuvant studies of one or two years (short-term) tamoxifen therapy have been sufficiently encouraging to promote additional applications for the drug. The long-term tamoxifen treatment protocols are in place in the United States, Britain, Sweden and France to answer the question of whether indefinite therapy will be an optimal strategy for stage I and II breast cancer.

However, it is the low incidence of side effects that has caused the current interest in the use of tamoxifen as a chemopreventative agent.

CHEMOPREVENTION OF BREAST CANCER WITH TAMOXIFEN

Approximately one third of patients with advanced breast cancer respond to endocrine therapy. It is believed that as breast cancer progresses, the disease is less hormone dependent, therefore, it could be argued that the early stages of carcinogenesis would be exclusively estrogen dependent. Certainly the evidence from carcinogen-induced rat mammary cancer support the essential role of hormones (Welsch, 1986).

On April 7, 1936, Lacassagne read a paper entitled "Hormonal Pathogenesis of Adenocarcinoma of the Breast" before the members of the American Association for Cancer Research in Boston. His work provided the link between hormones and the development of mouse mammary cancer. The paper concluded (paraphrased) that if one concedes that breast cancer is the consequence of a special hereditary sensitivity to the proliferative action of estrone then one might imagine the development of a therapeutic preventative. This could either be the use of a hormone that was antagonistic or perhaps an approach to facilitate the excretion of estrone thereby preventing its "stagnation" in the ducts of the breasts. Some fifty years later tamoxifen, an estrogen antagonist, is being considered as a chemopreventative.

The case for the use of tamoxifen is 1) a low toxicity in humans but efficacy as an adjuvant therapy, 2) the co-administration of tamoxifen and polycyclic hydrocarbons will prevent mammary carcinogenesis in the rat (Jordan, 1974, 1976), and 3) the administration of tamoxifen as a monoadjuvant therapy prevents the appearance of second-primary breast cancers in the contralateral breast (Cuzick and Baum, 1985).

Large clinical trials have been proposed to evaluate the efficacy of tamoxifen in both premenopausal or postmenopausal women. Women with a high risk of developing breast cancer would be recruited to participate but there is considerable debate about who should be considered to be at "high risk". There is a concern that the unrestricted use of tamoxifen in premenopausal women will potentially cause major perturbations of the endocrine system e.g., hyperstimulation of the ovaries. Similarly, the participants of trials must be carefully counciled about contraception. There are no data about tamoxifen being free of teratogenic potential in primates. To avoid these concerns a large population of postmenopausal women could be recruited. However, the application of prolonged tamoxifen therapy to this age group (>55 yrs) cannot be considered to be chemoprevention. The probability is that the drug would be controlling the appearance of pre-existing disease. The correct description of this approach would be chemosuppression (Jordan, 1986b). In fact, it should perhaps be stressed that the whole concept of hormonal chemoprevention may be invalid as a concept. There is currently no indication of when the carcinogenic insult occurs in humans. This is simple to define in the carcinogen-induced animal models i.e., at a time of early maturity. Clearly if this was found to be true for women e.g., smoking may cause the carcinogenic insult in 13-15 year old girls, then pharmacological chemopreventative agents would, almost certainly, be unacceptable for testing in this population.

Several toxicological issues must be addressed before tamoxifen can be considered for large scale clinical trials in older women. Estrogen appears to be physiologically important for women to maintain bone and to prevent the development of atherosclerosis. One could, therefore, take the position that long-term therapy with an antiestrogen might precipitate osteoporosis and predispose patients to coronary artery disease.

We are currently studying these problems at the University of Wisconsin Clinical Cancer Center with the Wisconsin Tamoxifen Study (Richard R. Love, M.D., Coordinator and Principal Investigator for the American Cancer Society grant). Patients with stage I breast cancer are being randomized to a two-arm study: tamoxifen 10 mg bid or placebo for two years. Periodic bone density determinations and blood lipid analyses will determine the action of tamoxifen upon these parameters. The study is double blind but the endocrinology of the postmenopausal women will be monitored 1) to study the impact of tamoxifen and 2) to ensure that patients are truely postmenopausal. Similarly, compliance will be established by the assay of tamoxifen and metabolites in all patients. We are concerned that patients suspecting they are in the placebo arm may attempt to obtain the drug from other sources.

We are optimistic that although the drug is classified as an antiestrogen which controls estrogen-stimulated growth, there may be sufficient estrogenic activity inherent in the molecule to facilitate other estrogenic processes. Preliminary studies in old rats demonstrate that tamoxifen does not precipitate osteoporosis in the intact animal, but retards the osteoporosis produced by ovariectomy (Jordan and Lindgren, unpublished observation). Similarly, the drug produces a whole range of distinct estrogen-like effects (Furr and Jordan, 1984) that may translate to estrogen-like effects on blood lipids. An earlier report in tamoxifen treated patients leads us to believe that the action of the drug may not be deleterious with regard to blood lipids (Rossner and Wallgren, 1984). Indeed, this result would re-inforce the hypothesis that the pharmacological effects of the drug can be completely different in different tissues. Thus, tamoxifen may prevent cell replication in breast tissue but produce estrogen-like actions elsewhere in the body.

However, some estrogen-like effects of tamoxifen may be of concern during prolonged therapy. Estrogens reduce the circulating levels of antithrombin III and put the patient at risk for thromboembolic disorders. Tamoxifen produces a decrease in circulating antithrombin III (Enck and Rios, 1984) but generally this is not to the level of clinical significance (lower than 70% of laboratory controls). Nevertheless, it would probably be unwise to treat women with a history of thromboembolic disorders.

SUMMARY AND CONCLUSIONS

The successful evaluation of tamoxifen as an antiestrogenic therapy for advanced breast cancer in the early 1970's, has resulted in its availability in more than 70 countries around the world. Currently the drug development process is focusing attention upon long-term adjuvant therapy and the future prospect of chemosuppression. Progress at this stage, however, must be cautious. Trials conducted using women with stage I disease have a high proportion of women who may never have a recurrence. At this point, the risk is justified because the toxicity of tamoxifen is low and disease recurrence is invariably very difficult or impossible to control. Future studies in the general population must be carefully weighed to ensure that the hazards do not exceed the benefits.

The pharmacology of tamoxifen seems to be a balance of estrogenic and antiestrogenic effects. Longer treatments with the drug must be carefully monitored. Uterine tissue should be examined to ensure that excessive stimulation does not occur. This is particularly true in the light of the recent report that a human uterine carcinoma, transplanted into athymic mice, can grow more rapidly during tamoxifen therapy (Satyaswaroop et al., 1984; Clark and Satyaswaroop, 1985). In fact, we have recently confirmed this observation in a collaborative study with Dr. Satyaswaroop. We have demonstrated that when athymic mice are transplanted in one axilla with an MCF-7 breast tumor and the human endometrial tumor in the other, tamoxifen causes the endometrial tumor to grow, but not the breast tumor. This again illustrates the target site specific effects of tamoxifen.

If, in the long run, the estrogenic side effects of tamoxifen are too severe then there is a case for the development of a non-estrogenic antiestrogen. Clearly this may provide benefit for short term (1-2 years) therapy and avoid any estrogen-like stimulation of tumor growth. Similarly, the concern about antithrombin III depression will be avoided. On the negative side, however, the concerns about atherosclerosis and osteoporosis will again have to be addressed with a new generation of agents.

ACKNOWLEDGEMENTS

This paper is dedicated to the memory of the late Dr. A.L. Walpole. He was a colleague, mentor and friend who died before he saw the broad application of his discovery. Throughout the development of this drug I would like to thank Drs. Brian Newbould, Barry Furr, A. (Sandy) H. Todd, and Mrs. Lois Trench-Hines for their help, support and encouragement to enable me to complete the studies described in this paper. The work during the past fifteen years has been supported by the Yorkshire Cancer Campaign (England), grants PO1-20432, P30-CA-14520, CA-32713 from the National Institutes of Health, grant PDT-290 from the American Cancer Society, and gift funds to the UWCCC from Imperial Chemical Industries PLC, England and Stuart Pharmaceuticals, Wilmington, Delaware.

REFERENCES

Baum M and other members of the Nolvadex Adjuvant Trial Organization (1985). Controlled trial of tamoxifen as single adjuvant agent in management of early breast cancer. Lancet i: 836-840.

Brown RR, Bain RR, Jordan VC (1983). Determination of tamoxifen and metabolites in human serum by high performance liquid chromatography with postcolumn fluorescence activation. J Chromatog 272: 351-358.

Clark CL, Satyaswaroop PG (1985). Photoaffinity labeling of the progesterone receptor from human endometrial carcinoma. Cancer Res 45: 5417-5420.

Cole MP, Jones CTA, Todd IDH (1971). A new antioestrogenic agent in late breast cancer. An early clinical appraisal of ICI 46,474. Br J Cancer 25: 270-275.

Cummings FJ, Gray R, Davis TE, Tormey DC, Harris JE, Falkson G, Arseneau J (1985). Adjuvant tamoxifen treatment of elderly women with stage II breast cancer. Annals Int Med 103: 324-328.

Cuzik J, Baum M (1985). Tamoxifen and the contralateral breast. Lancet ii: 282.

Delozier T, Julien J-P, Juret P, Veyret C, Couette J-E, Graic Y, Ollivier J-M, deRanieri E (1986). Adjuvant tamoxifen in postmenopausal breast cancer: Preliminary results of a randomized trial. Breast Cancer Res Treat 7: 105-110.

Dnistrian AM, Schwartz MK, Fracchia AA, Kaufman RJ, Hakes JB, Carrie VE (1983). Endocrine consequences of CMF adjuvant therapy in premenopausal and postmenopausal breast cancer patients. Cancer 51: 803-807.

Dodds EC, Golberg L, Lawson W, Robinson R (1938a). Oestrogenic activity of certain synthetic compounds. Nature 141: 247-248.

Dodds EC, Lawson W, Noble RL (1938b). Biological effects of the synthetic oestrogenic substance 4:4'-dihydroxy-α:β diethylstilbene. Lancet i: 1384-1391.

Enck RE, Rios CN (1984) Tamoxifen treatment of metastatic breast cancer and antithrombin III levels. Cancer 53: 2607-2609.

Fisher B, Redmond C, Brown A, Wickerman DL, Wolmark N, Allegra J, Esher G, Lippman ME, Savlov E, Wittliff JL, Fisher ER (1984). Influence of tumor estrogen and progesterone receptor levels on the response to tamoxifen and chemotherapy in primary breast cancer. J Clin Oncol 1: 227-241.

Fisher B, Redmond C, Brown N, Wolmark N, Wittliff J, Fisher ER, Plotkin D, Bowman D, Sachs S, Wolter J, Frelick R, Desser R (1981). Treatment of primary breast cancer with chemotherapy and tamoxifen. New Engl J Med 305: 1-6.

Furr BJA, Jordan VC (1984). The pharmacology and clinical uses of tamoxifen. Pharm Ther 25: 127-205.

Gottardis MM, Robinson SP, Jordan VC (1985). Control of estrogen responsive human breast cancer cell lines in athymic mice by long-term tamoxifen therapy. Breast Cancer Res Treat 6: 173.

Groom GV, Griffiths K (1976). Effect of the antioestrogen tamoxifen on plasma levels of luteinizing hormone, follicle-stimulating hormone, prolactin, oestradiol and progesterone in normal premenopausal women. J Endocrinol 70: 421-428.

Harper MJK, Walpole AL (1966). Contrasting endocrine activities of cis and trans isomers in a series of substituted triphenylethylenes. Nature (Lond) 212: 87.

Harper MJK, Walpole AL (1967a). A new derivative of triphenylethylene: effect on implantation and mode of action in rats. J Reprod Fert 13: 101-119.

Harper MJK, Walpole AL (1967b). Mode of action of ICI 46,474 in preventing implantation in rats. J Endocr 37: 83-92.

Haddow A, Watkinson JM, Paterson E, Koller PC (1944). Influence of synthetic oestrogens upon advanced malignant disease. Br Med J ii: 393-398.

Jordan VC (1974). Antitumour activity of the antioestrogen ICI 46,474 (tamoxifen) in the dimethylbenzanthracene (DMBA)-induced rat mammary carcinoma model. J Steroid Biochem 5: 354.
Jordan VC (1975). The effects of tamoxifen in the dimethylbenzanthracene (DMBA)-induced rat mammary carcinoma model. In "Proceedings of a Symposium on the Hormonal Control of Breast Cancer," Macclesfield: ICI Ltd Pharmaceuticals Division, pp 11-17.
Jordan VC (1976a). Antiestrogenic and antitumor properties of tamoxifen in laboratory animals. Cancer Treat Rep 60: 1409-1419.
Jordan VC (1976b). Effect of tamoxifen (ICI 46,474) on initiation and growth of DMBA-induced rat mammary carcinoma. Eur J Cancer 12: 419-424.
Jordan VC (1978). Use of the DMBA-induced rat mammary carcinoma system for the evaluation of tamoxifen as a potential adjuvant therapy. Reviews on Endocrine Rel Cancer (Oct suppl) ICI Ltd Pharmaceuticals Division, pp 49-55.
Jordan VC (1982). Laboratory models of hormone-dependent cancer. In Furr BJA (ed): "Clinics in Oncology," Philadelphia: W.B. Saunders Co., pp 21-40.
Jordan VC (1984). Biochemical pharmacology of antiestrogen action. Pharm Rev 36: 245-276.
Jordan VC (ed) (1986a). "Estrogen/Antiestrogen Action and Breast Cancer Therapy." University of Wisconsin Press, Madison.
Jordan VC (1986b). Tamoxifen prophylaxis: Prevention is better than cure - prevention is cure? In Cavalli F (ed): "Endocrine Therapy of Breast Cancer: Strategies and Future Directions," Heidelberg: Springer Verlag, pp 117-120.
Jordan VC, Allen KE (1980). Evaluation of the antitumour activity of the non-steroidal antioestrogen monohydroxytamoxifen in the DMBA-induced rat mammary carcinoma model. Eur J Cancer 16: 239-251.
Jordan VC, Allen KE, Dix CJ (1980). The pharmacology of tamoxifen in laboratory animals. Cancer Treat Rep 64: 745-759.
Jordan VC, Bain RR, Brown RR, Gosden B, Santos MA (1983). Determination and pharmacology of a new hydroxylated metabolite of tamoxifen observed in patient sera during therapy for advanced breast cancer. Cancer Res 43: 1446-1450.

Jordan VC, Dix CJ, Allen KE (1979). The effectiveness of long term tamoxifen treatment in a laboratory model for adjuvant hormone therapy of breast cancer. In Salmon SE, Jone SE (eds): "Adjuvant Therapy of Cancer II," New York: Grune & Stratton, pp 19-26.

Jordan VC, Dowse LJ (1976). Tamoxifen as an antitumour agent: Effect on oestrogen binding. J Endocr 68: 297-303.

Jordan VC, Fritz NF, Gottardis MM (1987). Strategies for breast cancer therapy with antiestrogens. J Steroid Biochem (in press).

Jordan VC, Fritz NF, Tormey DC (1987). Endocrine effects of adjuvant chemotherapy and long-term tamoxifen administration on node positive patients with breast cancer. Cancer Res 47: 624-630.

Jordan VC, Jaspan S (1976). Tamoxifen as an antitumor agent: Oestrogen binding as a predictive test for tumor response. J Endocr 68: 453-460.

Jordan VC, Koerner S (1976). Tamoxifen as an antitumour agent: Role of oestradiol and prolactin. J Endocr 68: 305-311.

Jordan VC, Mirecki D, Gottardis MM (1984). Continuous tamoxifen therapy prevents the appearance of mammary tumors in a laboratory model of adjuvant therapy. In Salmon SE, Jones SE (eds): "Adjuvant Therapy of Cancer IV," New York: Grune & Stratton, pp 27-33.

Jordan VC, Robinson SP (1987). Species specific pharmacology of antiestrogens: Role of metabolism. Fed Proc. 46: 1870-1874.

Lacassagne A (1936). Hormonal pathogenesis of adenocarcinoma of the breast. Am J Cancer 14: 217-228.

Lippman ME, Bolan G (1975). Oestrogen-responsive human breast cancer in long term tissue culture. Nature (Lond) 256: 592-595.

Ludwig Breast Cancer Study Group (1984). Randomized trial of chemoendocrine therapy and mastectomy alone in post menopausal patients with operable breast cancer and axillary node metastases. Lancet i: 1256-1260.

Lyman SD, Jordan VC (1985). Metabolism of tamoxifen and its uterotrophic activity. Biochem Pharmacol 34: 2787-2794.

Manni A, Pearson OH (1980). Antioestrogen-induced remissions in premenopausal women with stage IV breast cancer: Effects on ovarian function. Cancer Treat Rep 64: 779-785.

Nicholson RI, Davies P, Griffiths K (1977a). Effect of oestradiol-17β and tamoxifen on nuclear oestradiol-17β receptors in DMBA-induced rat mammary tumours. Eur J Cancer 13: 201-208.

Nicholson RI, Davies P, Griffiths K (1977b). Early increases in ribonucleic acid polymerase activities of dimethylbenzanthracene-induced mammary tumour nuclei in response to oestradiol-17β and tamoxifen. J Endocr 73: 135-142.

Nicholson RI, Golder MP (1975). The effect of synthetic antioestrogens on the growth and biochemistry of rat mammary tumours. Eur J Cancer 11: 571-579.

Osborne CK, Boldt DA, Clark GM, Trent, JM (1983). Effects of tamoxifen on human breast cancer cell cycle kinetics: Accumulation of cells in early G_1 phase. Cancer Res 43: 3583-3585.

Osborne CK, Hobbs K, Clark CM (1985). Effect of estrogens and antiestrogens on growth of human breast cancer cells in athymic nude mice. Cancer Res 45: 584-590.

Pritchard KI, Meakin JW, Boyd NF, Ambuo U, DeBoer G, Dembo AJ, Paterson AHG, Sutherland DJA, Wilkinson RH, Bassett AA, Evans WK, Beale FA, Clark RM, Keane TJ (1984). A randomized trial of adjuvant tamoxifen in postmenopausal women with axillary node positive breast cancer. In Jones SE, Salmon SE (eds): "Adjuvant Therapy of Cancer IV," New York: Grune & Stratton, pp 339-347.

Ribeiro G, Palmer MK (1983). Adjuvant tamoxifen for operable carcinoma of the breast: Report of the clinical trial by the Christie Hospital and Holt Radium Institute. Br Med J 286: 827-830.

Rose C, Thorpe SM, Anderson KW, Pedersen BV, Mouridsen HT, Blicher-Toft M, Rasmussen BB (1985). Beneficial effect of adjuvant tamoxifen therapy in primary breast cancer patients with high oestrogen receptor values. Lancet i: 16-18.

Rossner S, Wallgren A (1984). Serum lipoproteins and proteins after breast cancer surgery and effects of tamoxifen. Atherosclerosis 52: 339-346.

Satyaswaroop PG, Zaino RJ, Mortel R (1984). Estrogen-like effects of tamoxifen on human endometrial carcinoma transplanted into nude mice. Cancer Res 44: 4006-4010.

Sawka CA, Pritchard KE, Paterson AHG, Sutherland DJA, Thomsen DB, Shelley WE, Myers RE, Mobbs BG, Malkin A, Meakin JW (1986). Role and mechanism of action of tamoxifen in premenopausal women with metastatic breast cancer. Cancer Res 46: 3152-3156.

Shafie SM, Grantham FH (1980). Role of hormones in the growth and regression of human breast cancer cells (MCF-7) transplanted into athymic nude mice. J Natl Cancer Inst 67: 51-56.

Sherman BM, Chapler FK, Crickard K, Wycoff D (1979). Endocrine consequences of continuous antiestrogen therapy in premenopausal women. J Clin Invest 64: 398-404.

Skidmore JR, Walpole AL, Woodburn J (1972). Effect of some triphenylethylenes on oestradiol binding in vitro to macromolecules from uterus and anterior pituitary. J Endocr 52: 289-298.

Soule HD, McGrath CM (1980). Estrogen responsive proliferation of clonal human breast carcinoma cells in athymic mice. Cancer Let 10: 177-181.

Stewart HJ, Prescott R (1985). Adjuvant tamoxifen therapy and receptor levels. Lancet i: 573.

Sutherland RL, Green MD, Hall RE, Reddel RR, Taylor IW (1983). Tamoxifen induces accumulation of MCF-7 human mammary carcinoma cells in the G_0/G_1 phase of the cell cycle. Eur J Cancer Clin Oncol 19: 615-621.

Sutherland RL, Jordan VC (eds) (1981). "Non-Steroidal Antioestrogens: Molecular Pharmacology and Antitumour Actions." Academic Press, Sydney.

Sutherland RL, Murphy LC (1982). Mechanism of oestrogen antagonism by nonsteroidal antioestrogens. Mol Cell Endocrinol 25: 5-23.

Sutherland RL, Watts CKW, Ruenitz PC (1986). Definition of two distinct mechanisms of action of antiestrogens on human breast cancer proliferation using hydroxytriphenylethylenes with high affinity for the estrogen receptor. Biochem Biophys Res Comm 140: 523-529.

Tormey DC, Jordan VC (1984). Long-term adjuvant therapy in node positive breast cancer - a metabolic and pilot clinical study. Breast Cancer Res Treat 4: 297-302.

Tormey DC, Rasmussen P, Jordan VC (198). Long-term tamoxifen therapy. Breast Cancer Res Treat (letter to the editor) in press.

Wallgren A, Baral E, Carsten J, Friberg S, Glas U, Gjalmar M-L, Kaigas M, Nordenskjold B, Skoog L, Theve N-O, Wiking N F(1984). Should adjuvant tamoxifen be given for several years in breast cancer. In Jones SE, Salmon SE (eds): "Adjuvant Therapy of Cancer IV," New York: Grune & Stratton, pp 331-337.

Walpole AL, Paterson E (1949). Synthetic oestrogens in mammary cancer. Lancet ii: 783-786.

Welsch CW (1985). Host factors affecting the growth of carcinogen-induced rat mammary carcinomas: A review and tribute to Charles Brenton Huggins. Cancer Res 45: 3415-3443.

Wilson AJ, Tehrani F, Baum M (1982). Adjuvant tamoxifen therapy for early breast cancer: An experimental study with reference to oestrogen and progesterone receptors. Br J Surg 69: 121-125.

ESTROGEN RECEPTOR IN HUMAN PROSTATE: ASSAY CONDITIONS FOR QUANTITATION

S. Ganesan, N. Bashirelahi and J.D. Young, Jr.
Department of Biochemistry, Dental School and the Department of Surgery, Division of Urology, School of Medicine, University of Maryland, Baltimore, MD 21201, USA

INTRODUCTION

Benign prostatic hypertrophy (BPH) is seen in 75% of men over the age of sixty. While little is known of the etiology of the disease, it appears to have an endocrine origin since early castration prevents the occurrence of BPH (Wilson, 1980). Although the growth and functional activities of the prostate are primarily under androgenic control, there is a considerable body of evidence which suggests that estrogens are also involved in the control of this gland (Wilson, 1980; DeKlerk et al., 1979; Habenicht et al., 1986; Stone et al., 1986).

Estrogens are known to act via intracellular receptors (Jensen et al., 1968; King and Greene, 1984; Welshons et al., 1984). Hence, if estrogens are to act on BPH tissue, it is necessary that the BPH tissue contain estrogen receptors. The presence of estrogen receptors in human prostate has, however, remained controversial (Bashirelahi et al., 1976, 1979; Ekman et al., 1979, 1983; Hawkins et al., 1975; Krieg et al., 1981; Wagner et al., 1975; de Voogt and Dignjan, 1980; Donnelly et al., 1983). Detection of steroid hormone receptors in human prostate has been difficult and has required delicate manipulation of the assay conditions. The prostate gland consists of four different anatomic regions of substantially different composition (McNeal, 1980). A heterogeneity in receptor distribution in human prostate has been reported (Bashirelahi et al., 1983). Estrogen receptor in BPH has been suggested to be localized predominantly in the stroma (Bashirelahi et al., 1979; Krieg et al., 1981).

The buffer system routinely used for estrogen receptor analysis consists of 10 mM Tris-HCl, 1.0 or 1.5 mM Na_4EDTA (ethylenediamine tetraacetate) and 1 mM DTT (dithiothreitol), with or without 0.1 or 1.0 mM PMSF (phenylmethylsulfonyl fluoride) (Donnelly et al., 1983; Ekman et al., 1983). This buffer system was originally developed for estrogen receptor analysis in calf and rat uteri (Clark and Peck, 1983), but has been routinely adapted for human prostate system as well. DTT has been shown to inhibit estradiol binding to Type II sites in rat uterus (Markaverich et al., 1981; Eriksson et al., 1978) and in rat prostate (Swaneck et al., 1982). EDTA is a known chelator of divalent cations and is routinely used in buffer systems designed to assay steroid receptors (Ekman et al., 1983). Molybdate has been widely used to stabilize steroid receptors in the non-transformed state (Leach et al., 1979; Nishigori and Toft, 1980). Protease inhibitors like PMSF have been routinely added to buffer systems used for receptor analysis in an effort to inhibit endogenous proteolytic activity (Donnelly et al., 1983; Ekman et al., 1983).

The controversy regarding the presence of estrogen receptors in human prostate might be due to the fact that optimum assay conditions have not been used. In view of the potential importance of estrogen receptors in human prostate, we have systematically reinvestigated the effect of buffer systems, reducing agents, EDTA, molybdate and PMSF on the binding of 17 β-estradiol to human benign prostatic cytosol.

EXPERIMENTAL

[^{3}H] 17 β -estradiol (95 Ci/mmol) was purchased from New England Nuclear (Boston, MA). PMSF was purchased from Bethesda Research Laboratories, Inc. (Gaithersburg, MD). All other chemicals and diethylstilbestrol (DES) were purchased from Sigma Chemical Co. (St. Louis, MO).

Buffers

Tris-HCl (10 mM) was adjusted to pH 7.4 at room temperature. Potassium phosphate (17 mM) was also adjusted to pH 7.4 at room temperature. DTT, EDTA, sodium molybdate and PMSF were added to assay tubes to give the final concentrations indicated in the experiments. Stock solutions of DTT, EDTA and molybdate were prepared in double distilled water. A 0.1 M stock solution of PMSF was prepared in 95% ethanol.

Tissue Handling and Storage

Human prostate tissue was obtained from the University of Maryland and Sinai Hospitals, Baltimore, immediately following surgery in containers with ice-cold normal saline solution. The specimens were cut into small pieces, weighed, placed in plastic vials, quickly frozen in liquid nitrogen and stored at $-70^{\circ}C$ until further processing. Prostate samples were obtained by open enucleations and were histologically identified as glandular and/or stromal hyperplasia. Prostate tissue from ten different patients was used in the study.

Preparation of Cytosol

Cytosol was prepared as described (Bevins and Bashirelahi, 1980). The protein concentration was determined by the method of Lowry et al., 1951.

Binding Assays

Cytosol was diluted with buffer to a protein concentration of 2-3 mg/ml and incubated overnight with [^{3}H] 17 β-estradiol (0.5-5.0 nM) at $0-4^{\circ}C$ in the presence or absence of a 1,000 fold excess of DES. Free steroid was removed by incubating with an equal volume of a dextran-coated charcoal (DCC) suspension (0.5% norite A, 0.005% dextran in buffer, pH 7.4 at room temperature) for 10 min. The DCC was pelleted by centrifugation at 5,000g for 10 min. Aliquots of the supernatant were then taken for measurement of radioactivity.

Glycerol Density Gradient Centrifugation

Linear 10-30% glycerol gradients were prepared in 5 ml polyallomer tubes using a locally constructed gradient former and a peristaltic pump (Technicon Instrument Corp., Tarrytown, New York). Cytosol (0.5 ml) was incubated with 10 nM [^{3}H] 17 β-estradiol for 16-20 hrs at 0-4°C to determine the total bound radioactivity. A parallel incubation with a 1,000-fold excess of unlabelled DES was carried out to determine non-specific bound radioactivity. The incubations were ended by treating them with DCC pellets to remove the free steroid. For this purpose, 0.5 ml of 0.5% DCC suspension was pelleted by centrifugation at 5,000g for 10 min. The incubations were pipetted onto the DCC pellets and briefly mixed with a vortex mixer. After a 10 min incubation in an ice-bath, the DCC was pelleted again as before. A portion (0.3 ml) of the supernatant was layered on the gradients and centrifuged in a Beckman swinging bucket rotor (SW 50) for 16 hrs at 40,000 rpm. Fractions were collected by inserting a thin steel tube to the bottom of the gradient and removing the contents by a peristaltic pump. Three drop fractions (about 0.15 ml) were collected into scintillation vials using a LKB fraction collector (Stockholm, Sweden). Radioactivity was measured as described below. Molecular markers, to determine "S" values (γ globulin 7·12 S, and bovine serum albumin 4.6 S) were added to parallel gradients and processed as indicated above. After fractionation, 1 ml of distilled water was added to each fraction and the absorbance at 280 nm of each fraction was determined.

Specificity of Binding

Cytosol (0.2 ml) was incubated with 1 nM [^{3}H] 17 β-estradiol in the absence or presence of 10-, 100-, 1,0000- or 10,000-fold excess of unlabelled competitors. The competitors used in the experiments were DES, 17 β-estradiol, promegestone (R5020), dihydrotestosterone (DHT) and dexamethasone. The tubes were incubated overnight at 0-4°C. The incubations were ended by treatment with DCC as described earlier for binding assays. Aliquots were removed for measurement of radioactivity.

Measurement of Radioactivity

Scintillation cocktail was prepared and radioactivity was measured as described (Patterson and Green, 1965).

Statistical Analysis

The steepest portion of the Scatchard plots (Scatchard, 1949) and the lines for the double reciprocal plots were drawn by linear regression analysis. The effect of molybdate was compared against control by a paired t-test (Table 1). Multiple group means were compared by analysis of variance (AnoVa). When the AnoVa test indicated that at least one of the groups was different, Dunnett's analysis (Dunnett, 1964) was used to compare the test groups against the control.

RESULTS

When specific binding (B_{max}) expressed as femtomoles/mg protein and K_D (equilibrium dissociation constant) values were compared between cytosols prepared with tris or phosphate buffers, the differences were not statistically significant (data not shown). Only phosphate buffer was used in further experiments.

Two classes of binding (high affinity-Type I, low affinity-Type II) could be detected in both saturation curves and Scatchard plots (data not shown). However, the quantity of Type II binding present varied between the assays. The initial steepest protion of the Scatchard plot was extrapolated to determine the B_{max} (maximum number of binding sites) and K_D. Although this method leads to some overestimation of Type I sites, it was done on all assays for comparative purposes.

[^{3}H]-estradiol binding sites were titrated with 0-1 mM DTT and the data were analyzed using Scatchard plots. These concentrations of DTT are routinely used in receptor analysis and hence were chosen for this investigation. The treatment groups were compared against control by Dunnett's analysis (Dunnett, 1964). The B_{max} of all the treatment groups was significantly lower than that of control (data not shown). The K_D values were, however, not significantly different (data not shown). The B_{max} obtained in the presence of 1 mM DTT was 90 +/- 35 fmoles/mg protein with a K_D of 2.7 +/- 1.2 nM.

Scatchard analysis was performed in the presence of 0, 1 and 10 mM EDTA. The data were statistically evaluated as described for DTT. The B_{max} obtained in the presence of 1 mM EDTA was not significantly different from that of control (Table 1). However, the B_{max} obtained in the presence of 10 mM EDTA was significantly higher than that of control (Table 1). EDTA (10 mM) stimulated estradiol binding even in the presence of 1 mM DTT (Figure 1). Similar results were also obtained by glycerol density gradient centrifugation (data not shown).

Table 1. Effect of ethylenediamine tetraacetate (EDTA) and molybdate on estradiol binding in human benign prostatic hyperplasia.

	EDTA (mM)			Molybdate (mM)	
	0	1.0	10.0	0	20
B_{max} fmol/mg protein	130 +/- 73	81 +/- 49 NS	203 +/- 121 *	125 +/- 73	69 +/- 44 **
K_D nM	2.5 +/- 1.2	1.5 +/ 1.0 NS	3.6 +/- 1.8 NS	3.2 +/- 1.4	2.4 +/- 1.2 ***

Saturation analysis was done in the absence (control) or presence of EDTA and molybdate. [^{3}H] 17 β-estradiol (0.5-5.0 nM) was incubated with cytosol at 0-4°C for 16-20 hrs in the absence or presence of a 1,000-fold excess of DES. Free steroid was removed with DCC. B_{max} and K_D were obtained from Scatchard plots. The number of experiments done were 7 for EDTA and 9 for molybdate. * $p < 0.05$; ** $p < 0.005$; *** $p < 0.01$; NS - not significant.

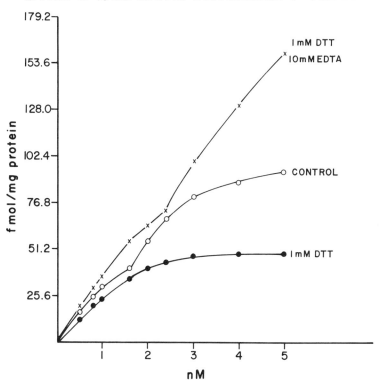

Figure 1. The effect of 10 mM EDTA on estradiol binding was analyzed in the presence of 1 mM DTT. Cytosol was incubated with [^{3}H] 17 β-estradiol (0.5-5.0 nM) at 0-4°C for 16-20 hrs in the presence or absence of a 1,000-fold excess DES. Free steroid was removed with DCC. Only specific binding was plotted for comparison.

The effect of 20 mM molybdate on [^{3}H] 17 β-estradiol binding was studied by Scatchard plots and glycerol density gradient centrifugation. The B_{max} and K_D values obtained in the presence of 20 mM molybdate were significantly lower than that of control

(Table 1). Glycerol density gradient analysis revealed a major specific estradiol binding peak in the 4-5 S region which was inhibited in the presence of 20 mM molybdate (Figure 2).

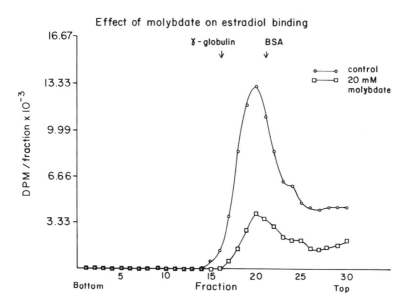

Figure 2. The effect of 20 mM molybdate on estradiol binding was analyzed by glycerol density gradient centrifugation. Cytosol (0.5 ml) was incubated with 10 mM [^{3}H] 17 β -estradiol at 0-4°C for 16-20 hrs. A parallel incubation was carried out with a 1,000-fold excess of unlabelled DES. The incubations were ended by treating them with DCC pellets. An aliquot (0.3 ml) was loaded onto a linear 10-30% glycerol gradient and processed as indicated in the experimental section. BSA and γ globulin were used as external markers. Only specific binding was plotted for comparison. The cytosol preparation had a protein concentration of 8.0 mg/ml.

Scatchard plots and glycerol density gradients were also used to analyze the effect of PMSF on estradiol binding. The concentrations of PMSF used were 0.1, 1.0 and 2.0 mM. PMSF (0.1 and 1.0 mM) is routinely used (Donnelly et al., 1983; Ekman et al., 1983) while 2.0 mM PMSF was arbitrarily chosen. The B_{max} values obtained in the presence of PMSF were significantly lower than that of control (Table 2). The K_D values obtained in the presence of PMSF were lower but not significantly different from the control values (Table 2). Double-reciprocal plots of the data indicated that the nature of the inhibition is uncompetitive (Figure 3). Glycerol density gradients also revealed that PMSF inhibited specific estradiol binding (data not shown).

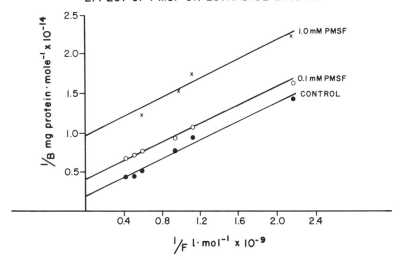

Figure 3. The effect of varying concentrations of PMSF on estradiol binding was analyzed by double reciprocal plots. Cytosol was incubated with 0.5-5.0 nM [^{3}H] 17 β-estradiol at 0-4°C for 16-20 hrs in the absence or presence of PMSF. Parallel incubations were carried out with a 1,000-fold excess of unlabelled DES. The bound steroid was determined by a DCC assay. ^{1}/Bound was plotted against ^{1}/Free. Lines were drawn by linear regression analysis.

Only estradiol and DES showed competition to [^{3}H] 17 β-estradiol (data not shown). All other compounds showed little or no competition. These results clearly demonstrated that the binding sites were highly specific for 17 β-estradiol.

The B_{max} (Type I binding) for all control experiments was 84 +/- 69 fmoles/mg protein with a K_D of 2.5 +/- 1.7 nM.

Table 2. Effect of phenylmethylsulfonyl fluoride (PMSF) on estradiol binding in human benign prostatic hyperplasia.

	PMSF (mM)			
	0	0.1	1.0	2.0
B_{max} fmol/mg protein	59 +/- 34	26 +/- 5 *	16 +/- 8 *	10 +/- 11 *
K_D nM	2.2 +/- 2.2	1.0 +/- 0.4 NS	0.8 +/- 0.5 NS	0.5 +/- 0.5 NS

Saturation analysis of estradiol binding was done in the absence (control) or presence of PMSF. Cytosol was incubated with [^{3}H] 17 β-estradiol (0.5-5.0 nM) at 0-4°C for 16-20 hrs in the absence or presence of a 1,000-fold excess unlabelled DES. Free steroid was removed with DCC. B_{max} and K_D were obtained from Scatchard plots. The results of six experiments were compared. * $p < 0.01$; NS - not significant.

DISCUSSION

Recent studies have indicated that the estrogen receptor might be a phosphoprotein and that the estrogen binding capacity of the receptor might be inactivated by a nuclear phosphatase (Auricchio et al., 1984). If phosphatase were a major factor in modulating estrogen receptor binding, its activity can be inhibited by using phosphate buffers. However, in our studies, phosphate buffer did not significantly improve binding over that observed with Tris buffer.

Heterogeneity of estradiol binding sites has been reported in rat uterus (Markaverich et al., 1981; Clark et al., 1978; Eriksson et al., 1978), rat prostate (Swaneck et al., 1982) and human prostate (Ekman et al., 1983). Curvilinear Scatchard plots have been reported in the cytosol (Eriksson et al., 1978; Ekman et al., 1983) and hooked Scatchard plots have been reported in the nuclei (Eriksson et al., 1978; Swaneck et al., 1982). Our results were similar to those reported for nuclear estrogen binding. A recent study has suggested that the presence of low concentrations of Type II (low affinity) binding results in a curvilinear Scatchard plot, whereas a high concentration of Type II binding results in a hooked Scatchard plot (Sheehan et al., 1986). Reducing agents like DTT have been shown to inhibit Type II binding (Markaverich et al., 1981; Swaneck et al., 1982). Our results confirm earlier observations that DTT inhibits estradiol binding to Type II sites. The mean value for B_{max} obtained in the presence of 1 mM DTT, 90 +/- 35 fmoles/mg protein, was comparable to the mean value for all control experiments, 84 +/- 68 fmoles/mg protein. The K_D values obtained in the presence of 1 mM DTT, 2.7 +/- 1.2 nM, were also comparable to the K_D values for all control experiments, 2.5 +/- 1.7 nM. This indicates that the K_D values reported in this study are true K_D values.

EDTA is a known chelator of divalent cations and is routinely added to buffer systems used for steroid receptor analysis (Bashirelahi et al., 1976; Ekman et al., 1983). The presence of high concentrations of Zn^{++} in human prostate has been established (Bertrans and Vladesco, 1921; Gyorkey et al., 1967). The effect of Zn^{++} on 17 β-estradiol and progesterone binding proteins in human endometrial and uterine cytosols (Habib et al., 1980; Sanborn et al., 1971) and androgen binding protein in human prostate (Habib and Stitch, 1975) have been studied. Recent studies have suggested that Zn^{++} might be involved in maintaining the active tertiary structure of the DNA binding domains of steroid receptors (Jeltsch et al., 1986). Chelating agents like EDTA could alter the free metal-ion concentration and thus influence steroid-receptor interactions. High concentrations of EDTA (10-50 mM) have been reported to inhibit progesterone receptor binding in human prostate (Imai et al., 1984). Our results indicate that there was no statistical difference in B_{max} between control and 1 mM EDTA. The reason for the enhancement of binding observed in the presence of 10 mM EDTA is not clear and requires further investigation.

Molybdate has been widely used to stabilize receptors in the non-transformed state (Leach et al., 1979; Nishigori and Toft, 1980). The mechanism of action of molybdate has been attributed to a direct interaction with the receptor (Nishigori and Toft, 1980; Grody et al., 1982) and also to its inhibitory action on phosphatases (Dipietro and Zengerie, 1967; Paigen, 1958). Stabilization of the 8S aggregated form of the progesterone receptor with molybdate in human BPH has been reported (Bevins and Bashirelahi, 1980). Our results indicate that 20 mM molybdate significantly inhibited estradiol binding in human BPH (Table 1, Figure 2). Earlier studies which reported negligible estrogen receptor binding in human BPH cytosol had used 25 mM molybdate (Ekman et al., 1983) and 40 mM molybdate (Robel et al., 1985). Our studies suggest that use of high concentrations of molybdate (20 mM or higher) could lead to inhibition of estrogen receptor binding and false negative results in human prostate.

One of the problems encountered with receptor analysis is the presence of endogenous proteolytic activity. Protease inhibitors like PMSF have often been added to buffer systems used for preparing cytosol in an effort to inhibit endogenous proteolytic activity (Donnelly et al., 1983; Ekman et al., 1983; Robel et al., 1985). The effectiveness of the protease inhibitors has, however, been controversial (Birnbaumer et al., 1983; Sherman et al., 1978). PMSF is a serine protease inhibitor that acts by sulfonylation of a serine hydroxyl group at the active site of the enzyme (Myers and Kemp, 1954). Contrary to stabilizing or enhancing estrogen receptor binding, PMSF actually inhibited estradiol binding at all concentrations used in this study. Earlier studies had indicated that the binding of steroid hormones (Baker et al., 1978, 1980; Baker and Kimlinger, 1982) and nerve growth factor (Stach et al., 1986) to their respective receptors were inhibited by PMSF. Another serine protease inhibitor, diisopropylfluoro phosphate (DIFP), has been shown to bind to partially purified estrogen receptor from calf-uterus (Puca et al., 1986). The nature of the interaction between PMSF and the estrogen receptor is, however, not clear and requires further investigation.

The heterogeneity of cell types (McNeal, 1980), the heterogeneity of steroid receptor binding (Bashirelahi et al., 1979) and the heterogeneity of estradiol binding sites (Type I and Type II) have complicated estrogen receptor measurements in human prostate. In the present study, no attempt was made to separate the tissue into histologically different regions. This has resulted in a wide range of values of Type I estradiol binding.

It is well known that treatment with estrogen causes regression of prostate cancer (Huggins and Hodges, 1941) but it is not clear whether this effect is mediated directly by the estrogens on the prostatic cell (Farnsworth, 1969) or indirectly by the suppression of the hypothalamic pituitary axis. It has been proposed that BPH in man may arise from an alteration in the ratio of androgen to estrogen in serum with aging (Wilson, 1980). BPH can be induced in castrated dogs by the simultaneous administration of 5 α androstane 3 β 17 β diol and 17 β -estradiol (DeKlerk et al., 1979). BPH can also be induced in castrated dogs by the simultaneous administration of androstenedione and 3 α, 17 β androstanediol and is characterized by hyperplasia of the epithelium (androgenic effects) and stroma (estrogenic effects). The estrogenic effects could be clearly antagonized by simultaneous treatment with the aromatase inhibitor 4-hydroxy androstenedione (Habenicht et al., 1986). The human BPH tissue has been shown to be able to aromatize androgens to estrogens (Stone et al., 1986). These studies suggest a role for estrogens in the etiology of BPH. However, earlier studies had shown only negligible concentrations of estrogen receptor in human BPH (Ekman et al., 1979, 1983; Robel et al., 1985). It has been argued that no significance can be attached to the low concentrations of estrogen receptor in human BPH. Thus, a role for estrogens in the etiology of BPH has remained elusive. Analysis of the assay protocols used in some of the earlier studies have revealed that PMSF and high concentrations of molybdate were used (Ekman et al., 1983; Robel et al., 1985). Based on our present study, the low binding reported in other studies was probably due to the use of PMSF and high concentrations of molybdate. In human prostate samples obtained by open adenoma enucleations and histologically identified to be glandular and/or stromal hyperplasia, we have observed a B_{max} (Type I binding) of 84 +/- 69 fmoles/mg protein (mean +/- SD) with a K_D of 2.5 +/- 1.7 nM. These values represent a mean of the control assays (in the absence of any added reagents) reported in this study. The values observed in this study are higher than those reported elsewhere (Ekman et al., 1983; Robel et al., 1985), even after allowing for possible overestimation.

Auf and Ghanadian, 1982, reported a K_D value of 2.9 +/- 2.6 nM for estradiol binding in human BPH. Sakai and Gorski, 1984, reported a K_D value of 3.2 nM for the native estrogen receptor from rat uterine cytosol. Simmen et al., 1984, reported a K_D value of 2.0 nM for Type I estradiol binding in chick liver nuclear matrix. Ekman et al., 1983, reported a K_D value of 0.1 +/- 0.09 nM for estradiol binding in human BPH, but their values were obtained in the presence of 1 mM PMSF and 25 mM molybdate. Our K_D value of 2.5 +/- 1.7 nM for control experiments (where Type II binding was not inhibited) is probably higher than the actual value due to contribution from Type II binding. Hence, our K_D values are comparable to those found in literature for the estrogen receptor.

The conditions used in this study (EDTA, PMSF, molybdate) have been found useful for estrogen receptor analysis in other systems. In the human prostate system, the estrogen receptors are present in a "male environment" (in the presence of large concentrations of testosterone) as opposed to other estrogen target tissues like the uterus and the mammary gland. It is also possible that there are as yet unidentified factors in prostate cytosol that affect estrogen receptor binding. The presence of high concentrations of Zn^{++} and other divalent cations in human prostate might also affect estrogen-receptor interactions. Further systematic studies are necessary to establish the reasons for the behaviour of estrogen receptor in human prostate.

Our studies indicate that optimum assay conditions might not have been used in earlier studies, leading to contradictory results. In view of the potential importance of estrogen receptors in human prostate, estrogen receptors must be carefully reevaluated using proper assay conditions. PMSF and high concentrations of molybdate must be omitted from buffer systems designed to assay estrogen receptors in human prostate. The nature of the interaction between EDTA and estrogen receptor is not clear and caution must be exercised in interpreting data obtained in its presence. We suggest that the estrogen receptors should be assayed using tris/phosphate buffer, with/without the addition of 1 mM DTT.

Acknowledgements: We are indebted to Dr. John Hebel, Professor, Epidemiology Dept., University of Maryland Medical School for his help with the statistical analysis. We thank Ms. Sheila McNair for typing this manuscript

REFERENCES

Auf G, Ghanadian R (1982). Characterization and measurement of cytoplasmic and nuclear estradiol 17 β receptor proteins in benign hypertrophied human prostate. J Endocr 93:305-317.

Auricchio F, Migliaccio A, Castoria G, Rotondi A, Lastoria S (1984). Direct evidence for the in vitro phosphorylation-dephosphorylation of the estradiol 17 β receptor. Role of Ca^{+2}- calmodulin in the activation of hormone binding sites. J Steroid Biochem 20:31-35.

Baker ME, Kimlinger WR (1982). Protease substrates inhibit binding of ^{3}H-R5020 to the G-fragment in chick oviduct cytosol. Biochem Biophys Res Commun 108:1067-1073.

Baker ME, Vaughn DA, Fanestil DD (1978). Inhibition by protease inhibitors of binding of adrenal and sex steroid hormones. J Supramol Struc 9:421-426.

Baker ME, Vaughn DA, Fanestil DD (1980). Competitive inhibition of dexamethasone binding to the glucocorticoid receptor in HTC cells by tryptophan methylester. J Steroid Biochem 13:993-995.

Bashirelahi N, O'Toole HH, Young JD (1976). A specific 17$_\beta$ estradiol receptor in human benign prostatic hypertrophy. Biochem Med 15:254-261.

Bashirelahi N, Young JD, Shida K, Yamanaka H, Ito Y, Harada M (1983). Androgen, estrogen and progesterone receptors in peripheral and central zones of human prostate with adenocarcinoma. Urology 21:530-535.

Bashirelahi N, Young JD, Sidh SM, Sanefuji H (1979). Androgen, estrogen and progestogen and their distribution in the epithelial and stromal cells of human prostate. In Schroeder FH, de Voogt HJ (eds): "Steroid Receptors, Metabolism and Prostatic Cancer", Amsterdam: Excerpta Medica, pp 240.

Bevins CL, Bashirelahi N (1980). Stabilization of 8S progesterone receptor from human prostate in the presence of molybdate ion. Cancer Res 40:2234-2239.

Bertrans G, Vladesco R (1921). Intervention probable du zinc dans les phenomenes de fecondation chez les animaux vertebres. Cr Seanc Soc Biol Fil 173:176-180.

Birnbaumer M, Schrader WT, O'Malley BW (1983). Photoaffinity labelling of the chick progesterone receptor proteins. J Biol Chem 255:1637-1644.

Clark JH, Hardin JW, Upchurch S, Erkisson H (1978). Heterogeneity of estrogen binding sites in the cytosol of rat uterus. J Biol Chem 253:7630-7634.

Clark JH, Peck EJ (1983). Steroid hormone receptors: basic principles and measurements. In Schrader WT, O'Malley BW (eds): "Laboratory Methods Manual for Hormone Action and Molecular Endocrinology", Texas: Dept. of Cell Biology, Baylor College of Medicine, pp 1-1 - 1-65.

DeKlerk DP, Coffey DS, Ewing LL, McDermott IR, Reiner WG, Robinson CH, Scott WW, Strandberg JD, Talalay P, Walsh PC, Wheaton LG, Zirkin BR (1979). Comparison

of spontaneous and experimentally induced canine prostatic hyperplasia. J Clin Invest 64:842-849.
deVoogt HJ, Dignjan P (1980). Is there a place for the assay of cytoplasmic steroid receptors in the endocrine treatment of prostatic cancer? In Schroeder FH, deVoogt JH (eds): "Steroid Receptors, Metabolism and Prostatic Cancer," Amsterdam: Excerpta Medica, pp 265.
Dipietro DL, Zengerie FS (1967). Separation and properties of three acid phosphatases from human placenta. J Biol Chem 242:3391-3396.
Donnelly BJ, Lakey WH, McBlain WA (1983). Estrogen receptor in human benign prostatic hyperplasia. J Urol 130:183-187.
Dunnett CW (1964). New tables for multiple comparisons with a control. Biometrics 20:482-491.
Ekman P, Barrack ER, Greene GL, Jensen EV, Walsh PC (1983). Estrogen receptors in human prostate: Evidence for multiple binding sites. J Clin Endocr Metab 57:166-176.
Ekman P, Snochowski M, Dahlberg E, Bression D, Hogberg B, Gustafsson JA (1979). Steroid receptor content from normal and hyperplastic human prostates. J Clin Endocr Metab 49:205-215.
Eriksson H, Upchurch S, Hardin JW, Peck EJ, Clark JH (1978). Heterogeneity of estrogen receptors in the cytosol and nuclei of rat uterus. Biochem Biophys Res Commun 81:1-7.
Farnsworth WE (1969). A direct effect of estrogens on prostatic metabolism of testosterone. Invest Urol 6:423-427.
Grody WW, Schrader WT, O'Malley BW (1982). Activation, transformation and submit structure of steroid hormone receptors. Endocr Rev 3:141-163.
Gyorkey F, Min KW, Huff JA, Gyorkey P (1967). Zinc and magnesium in human prostate gland: normal, hyperplastic and neoplastic. Cancer Res 27:1348-1353.
Habenicht V-F, Schwarz K, Schweikert H-V, Neumann F, ElEtreby MF (1986). Development of a model for the induction of estrogen-related prostatic hyperplasia in the dog and its response to the aromatase inhibitor 4-hydroxy-4-androstene-3,17, dione: preliminary results. Prostate 8:181-194.
Habib FK, Maddy SW, Stitch SR (1980). Zinc induced

changes in the progesterone binding properties of the human endometrium. Acta Endocr, Copenh 94:99-106.

Habib FK, Stitch SR (1975). The interrelationship of the metal and androgen binding proteins in normal and cancerous human prostatic tissue. Acta Endocr, Copenh 199:129.

Hawkins EF, Nijs M, Brassinne C, Tagnon HJ (1975). Steroid receptors in human prostate. I. Estradiol 17 β binding in benign prostatic hypertrophy. Steroids 26:458-469.

Huggins C, Hodges CV (1941). Studies on prostatic cancer I. Effect of castration, estrogen and androgen injection on serum phosphatases in metastatic carcinoma of prostate. Cancer Res 1:293-297.

Imai K, Shimuzu K, Yamanaka H (1984). Studies of progestin specific binding protein in the human prostate (I): studies on basic conditions for the accurate quantitative assay method. Nippon Naibunpi Gakkai Zasshi 60:950-963.

Jeltsch JM, Krozowski Z, Quirin-Stricker C, Gronemeyer H, Simpson RJ, Garnier JM, Krust A, Jacob F, Chambon P (1986). Cloning of the chicken progesterone receptor. Proc Natl Acad Sci USA 83:5424-5428.

Jensen EV, Suzuki T, Kawashima T, Stumpf WE, Jungblut P, DeSombre ER (1968). A two step mechanism for the interaction of estradiol with rat uterus. Proc Natl Acad Sci USA 59:632-638.

King WJ, Greene GL (1984). Monoclonal antibodies localize estrogen receptor in the nuclei of target cells. Nature 307:745-747.

Krieg M, Klotzl G, Kaufmann J, Voigt KD (1981). Stroma of human benign prostatic hyperplasia: preferential tissue for androgen metabolism and estrogen binding. Acta Endocr, Copenh 96:422-432.

Leach KL, Dahmer MK, Hammond ND, Sando JJ, Pratt WB (1979). Molybdate inhibition of glucocorticoid inactivation and transformation. J Biol Chem 254:11884-11890.

Lowry OH, Rosebrough NJ, Farr AL, Randall RJ (1951). Protein measurement with folin phenol reagent. J Biol Chem 19:265-275.

Markaverich BM, Williams M, Upchurch S, Clark JH (1981). Heterogeneity of nuclear estrogen binding sites in rat uterus: a simple method for the quantitation of type I and type II sites by ^{3}H estradiol exchange. Endocrinology 109:62-69.

McNeal JE (1980). The anatomic heterogeneity of the prostate. In Murphy GP (ed): "Models for Prostate Cancer", New York: Alan R. Liss, pp 149.

Myers DK, Kemp A (1954). Inhibition of esterases by the fluorides of organic acids. Nature 173:33-34.

Nishigori H, Toft D (1980). Inhibition of progesterone receptor activation by sodium molybdate. Biochemistry 19:77-83.

Paigen K (1958). The properties of particulate phosphoprotein phosphatase. J Biol Chem 233:388-394.

Patterson MS, Green RC (1965). Measurement of low energy beta-emitters in aqueous solution by liquid scintillation counting of emulsions. Analyt Chem 37:845-874.

Puca GA, Abbodaza C, Nigro V, Armetta I, Medici N, Molinari AM (1986). Estradiol receptor has proteolytic activity that is responsible for its own transformation. Proc Natl Acad Sci USA 83:5367-5371.

Robel P, Eychenne B, Blondeau J-P, Baulieu E-E, Hechter O (1985). Sex steroid receptors in normal and hyperplastic human prostate. Prostate 6:255-267.

Sakai D, Gorski J (1984). Reversible denaturation of estrogen receptor and estimation of polypeptide chain molecular weight. Endocrinology 115:2379-2383.

Sanborn BM, Rao BR, Korenman SG (1971). Interaction of 17 β-estradiol to its specific uterine receptor. Evidence for complex kinetic and equilibrium behaviour. Biochemistry 10:4955-4961.

Scatchard G (1949). The attraction of proteins for small molecules and ions. Ann NY Acad Sci 51:660-672.

Sheehan DM, Medlock KL, Lyttle CR (1986). Identification of uterine nuclear type II estrogen binding sites in estrogen treated rats. J Steroid Biochem 25:37-43.

Sherman MR, Pickering LA, Rollwagen FM, Miller LK (1978). Meroreceptors: proteolytic fragments of receptors containing the steroid binding site. Fedn Proc 37:167-173.

Simmen RCM, Means AR, Clark JH (1984). Estrogen modulation of nuclear matrix-associated steroid hormone binding. Endocrinology 115:1197-1202.

Stach RW, Stach BM, Ennulat DJ (1986). Phenylmethylsulfonyl fluoride (PMSF) inhibits nerve growth factor binding to the high affinity (Type I) nerve growth factor receptor. Biochem Biophys Res Commun 134:1000-1005.

Stone NN, Fair WR, Fishman J (1986). Estrogen formation in human prostatic tissue from patients with and without benign prostatic hyperplasia. Prostate 9:311-318.

Swaneck GW, Alvarez JM, Sufrin G (1982). Multiple species of estrogen binding sites in the nuclear fractions of rat prostate. Biochem Biophys Res Commun 196-1441-1447.

Wagner RK, Schulze KH, Jungblut PW (1975). Estrogen and androgen receptor in human prostate and prostatic tumor tissue. Acta Endocr, Suppl 193:52.

Welshons WV, Lieberman ME, Gorski J (1984). Nuclear localization of unoccupied estrogen receptors. Nature 307:747-749.

Wilson JD (1980). The pathogenesis of benign prostatic hyperplasia. Am J Med 68:745-747.

VALIDATION OF THE EXCHANGE ASSAY FOR THE MEASUREMENT OF ANDROGEN RECEPTORS IN HUMAN AND DOG PROSTATES.

Abdulmaged M. Traish, Donald F. Williams, Neil D. Hoffman and Herbert H. Wotiz.
Boston University School of Medicine, Department of Biochemistry, 80 E. Concord Street, Boston, MA 02118

ABSTRACT

This study describes an exchange assay for measurement of cytosolic and nuclear androgen receptors (AR) in human and dog prostates. Efficient replacement of endogenously bound ligand from the receptor with [^{3}H] mibolerone was achieved by incubation of cytosolic or nuclear fractions at 0°C for 72 h in the presence of 0.15M NaSCN and 15% sucrose. It was demonstrated that in the presence of the chaotropic salt rapid steroid dissociation took place at 0°C followed by [^{3}H] mibolerone binding; sucrose protected the AR from denaturation by NaSCN. Combination of these two reagents, therefore, allowed quantitative androgen exchange at 0°C. Receptors determined by this exchange procedure are specific for androgens, of high affinity (K_D 2-5nM), and sedimented on sucrose gradients as 4-4.6S entities. Use of [^{3}H] mibolerone minimized interference from plasma proteins and reduced nonspecific binding. This exchange assay has now been applied to quantitation of cytosolic and nuclear AR in small tissue samples (50 mg). Thus it is possible to measure AR in tissue samples obtained by needle biopsies and attempt to correlate receptor values to clinical response of prostatic cancer patients.

INTRODUCTION

While clinical correlation of estrogen and progesterone receptors with prognosis and therapeutic selection in breast cancer is well accepted, no such relationship between prostatic cancer and androgen receptors has been established as yet, despite numerous attempts. A possible reason for this lack may be the non-quantitativeness of the analytical procedures applied to measurement of AR. While several methods for the measurement of AR have been reported (Barrack et al 1983), (Carroll et al 1984), (Hicks and Walsh 1979), (Shain et al 1978), (Shain et al 1982) most of these procedures do not take into account the slow dissociation of endogenously bound hormone (half time >> 65h) at 0°C (Fiet and Muldoon 1983), (Hechter et al 1983), (Traish et al 1984), (Wilson and French 1976) and the rapid inactivation of androgen receptors at elevated temperatures (Fiet and Muldoon 1983), (Hechter et al 1983), (Olsson et al 1979), (Snochowski et al 1977). Thus, these methods measure only a relatively small fraction of total androgen receptors. Others have recognized the limitations described above and have developed alternative approaches to overcome them. Hechter et al(1983) have measured AR by exchange at 0°C for 96 h. These authors reported that only 50% of the receptor was detectable. They related this partial recovery of receptor to incomplete exchange, inactivation even at 0°C, and incomplete saturation at equilibrium.

Recently we have described an exchange assay for the measurement of AR in rat prostatic tissue (Traish et al 1985). Since the dog prostate is generally thought to be the most closely related and generally available animal model for the human prostate, it was thought important to validate this procedure in both species. We now wish to report on the validation of this method for quantification of cytosolic and nuclear AR in human and dog prostates, and applicability of this assay to small tissue samples to mimic those obtained by needle biopsy.

MATERIALS AND METHODS

Isotopes and Chemicals:

Radioinert and radiolabeled Mibolerone (7α, 17α-dimethyl-19-northestosterone, [^{3}H]DMNT, 82 Ci/mmol) were purchased from Amersham (Arlington Heights, IL). Methyltrienolone (R1881) was purchased from New England Nuclear Corp. (Boston, MA). All other steroids were obtained from Steraloids (Wilton, NH). Reagents were of analytical grade and were purchased from commercial sources.

Buffers and Solutions:

Buffer Tes/EGM: 20mM N-Tris-hydroxymethyl-methyl-2-aminoethane sulfonic acid, 1.5 mM ethylenediamine tetraacetic acid, disodium salt (EDTA), 10% (vol/vol) glycerol, 50mM molybdate, and sodium azide (0.02%) pH 7.4 at 0°C. Monothioglycerol (10mM) was added to the buffer immediately prior to use. Leupeptin (1mM) was included in the homogenization buffer and phenylmethyl sulfonylfluoride (PMSF, 0.5mM) was added during tissue homogenization to inhibit proteolysis.

Dog Prostate:

Dog prostates were obtained from Pel-Freeze Biologicals (Rogers, AR). Tissue was shipped on dry ice and stored at -70°C till use.

Human Tissue:

Human prostatic tissue specimens were obtained from patients undergoing prostatic surgery at University Hospital and Boston City Hospital. All specimens were immediately transported to the laboratory on dry ice and kept frozen in liquid nitrogen for no more than 6 weeks.

Preparation of Subcellular Fractions:

Unless otherwise stated all manipulations were carried out at 0-4°C. Frozen tissue samples were pulverized using a Thermovac tissue pulverizer which was precooled in liquid nitrogen. The tissue powder was quickly weighed and placed into a 50 ml thick-walled glass tube, buffer was added (4 ml of buffer/1 g tissue), and the tissue homogenized with a Polytron PT-10 using three 10-sec bursts with one minute intermittent cooling. This preparation was further homogenized in a Duall glass-glass homogenizer with a motor driven pestle. In the case of small samples (50 - 100 mg) tissue was homogenized in 10 volumes of buffer using glass-glass tissue homogenizers. The homogenate was then centrifuged at 1000 x g for 10 min at 2°C to isolate the crude nuclear fraction. The supernatant was recentrifuged at 100,000 x g for 45 min at 2°C to obtain the cytosol. The crude nuclear fraction was washed twice with buffer (Resuspension followed by centrifugation) and again homogenized in a teflon-glass homogenizer in order to obtain a uniform nuclear suspension.

Exchange of Endogenous Bound Hormone with [^{3}H]DMNT; Effect of NaSCN:

Aliquots of cytosol or nuclear suspension were dispensed into siliconized test tubes. An equal volume of buffer containing 30% sucrose, 0.3M NaSCN, and various concentrations of [^{3}H]DMNT (0.5-15nM) was added, mixed well and incubated at 0°C for 72 h. Parallel samples were incubated as above with [^{3}H]DMNT in the presence of 100 fold unlabeled 5α-dihydrotestosterone (DHT) to determine nonspecific binding (Traish et al 1986). DHT was used to displace [^{3}H]DMNT bound exclusively to androgen receptors. Unaccelerated exchange experiments were carried out for 24 or 72h at 0°-4°C but NaSCN was omitted from the incubation. Protein bound radioactivity was determined by hydroxylapatite (HAP) adsorption (Traish et al 1984). All data points are the means of triplicate determination for each sample.

Protein and DNA determinations were carried out as described previously (Traish et al 1985a). Analysis of AR on sucrose density gradients (SDG) containing 0.4M KCl and 50 mM molybdate was carried out as described elsewhere (Traish et al 1985a).

RESULTS

Effect of NaSCN on the Quantitation of Human Prostatic AR.

In order to determine the extent of ligand exchange with human prostatic AR obtainable with a conventional type of assay (Hicks and Walsh 1979),(Shain et al 1982), aliquots of cytosol were incubated at 0°C for 24 h with various amounts of [^{3}H]DMNT and 15% sucrose; non-specific binding was determined by incubation of cytosol under identical conditions in the presence of 5uM DHT. This concentration of unlabeled

DHT was found to be effective in displacing [^{3}H]DMNT binding to AR (Traish et al 1986). A total of 17 fmol of AR/mg protein could be detected (Fig. 1A). When the same cytosol

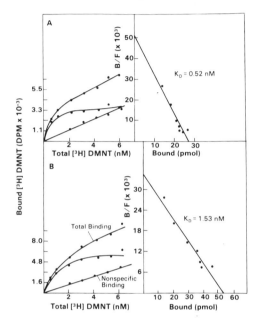

Fig. 1 Effect of NaSCN on Quantitation of Cytosolic AR in Human Prostate
Panel A Aliquots of human prostatic cytosol were incubated at 0°C for 24 h with an equal volume of buffer containing increasing concentrations of [^{3}H]DMNT and 30% sucrose. Bound radioactivity was assayed by HAP adsorption. Specific binding was plotted according to Scatchard.
Panel B Aliquots of the same cytosol were incubated at 0°C for 72 h with an equal volume of buffer containing increasing concentrations of [^{3}H]DMNT, 30% sucrose, and 0.3M NaSCN. Bound radioactivity was assayed with HAP and plotted according to Scatchard.

was analyzed by the new method about 35 fmol of AR/mg protein were measured (Fig. 1B). As has already been determined for rat prostatic receptors, NaSCN also decreased the affinity of human prostatic AR for androgen. This can be seen from the increase in K_D in the presence of the chaotropic salt. In similar experiments using the nuclear suspensions from these prostates [^{3}H]DMNT binding after 24 h at 0°C in the absence of NaSCN was significantly less when compared to that obtained with NaSCN/sucrose incubated for 72 h at 0°C (400 vs 850 fmol/mg DNA).

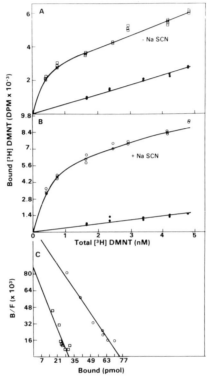

Fig. 2 Validation of the Exchange Assay for Human Cytosolic AR. Cytosol was incubated at 0°C for 24 h with 20nM unlabeled DHT. After the incubation free DHT was removed with DCC pellets without further dilution of the cytosol.

Panel A Aliquots of cytosol were then incubated as in Fig. 1A. Squares represent total binding, solid circles represent nonspecific binding.
Panel B Aliquots of cytosol were incubated as in Fig. 1B. Open circles represent total binding and solid circles represent nonspecific binding.
Panel C Scatchard analysis of the specific binding data presented in panel 2A and 2B. Open circles with NaSCN, squares without NaSCN.

Validation of the Exchange Assay Cytosolic and Nuclear AR from Human Prostate.

In order to verify that the experimental conditions described did indeed enhance the exchange of saturated receptor sites with radiolabeled ligand, human tissue cytosol and nuclear suspension were preincubated with 20nM of unlabeled DHT at 0°C for 24 h to saturate empty receptor sites with the natural ligand. Free hormone was removed from the cytosol with dextran coated charcoal (DCC), and from the nuclear suspension by three buffer washes. Aliquots of cytosol or nuclear suspensions were then re-incubated either at 0°C for 24 h (Fig. 2A) to mimic experimental conditions of the conventional assays (Barrack et al 1983), (Hicks and Walsh 1979) or at 0°C for 72 h with NaSCN and sucrose (Traish et al 1985b). As before, the data in Fig. 2 clearly demonstrate that in the same cytosol the number of androgen binding sites obtained in the presence of NaSCN (24 fmol of AR/mg protein) was significantly greater than that measured in the absence of NaSCN (10 fmol AR/mg protein). Similarly, increases in measurable AR from 432 fmol to 1040 fmol/mg DNA were obtained with the same nuclear suspensions (Fig. 3) in the presence of NaSCN.

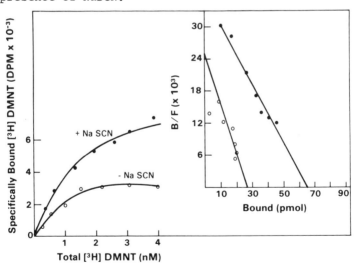

Fig. 3   Validation of the Exchange Assay with Nuclear AR.

Nuclear suspensions from human tissue were incubated with 20nM unlabeled DHT for 24 h at 0°C. Free DHT was removed by washing the nuclear pellets 3 times, using resuspension and centrifugation. Aliquots (62.5 ug DNA) of these nuclear suspensions were incubated exactly as described in Fig. 1A and 1B, and bound radioactivity was determined with HAP. Specific binding was determined and plotted according to Scatchard. Closed circles - with NaSCN, open circles - without NaSCN.

The data presented, thus far, compared binding of [^{3}H] mibolerone to cytosolic and nuclear AR at 0°C for 24 h with binding obtained at 0°C for 72 h in the presence of NaSCN. To establish that NaSCN accelerates exchange of ligands irrespective of the duration of the incubation the binding of [^{3}H] mibolerone to cytosolic and nuclear AR was determined at 0°C for 72 h in the absence or presence of NaSCN. As shown in Table 1, in the presence of NaSCN a significant increase in detectable AR is observed in both fractions. This increase is more pronounced with the nuclear fractions presumably because all nuclear bound AR is saturated with endogenous ligand and quantitive measurements of this receptor requires efficient exchange. The data clearly suggest that this assay is more quantitative than those methods which employ short term (24 h) incubation conditions at 0°-4°C in the absence of agents that accelerate the exchange of the ligand.

MEASUREMENT OF AR IN HUMAN PROSTATE BY EXCHANGE ASSAY

AR fmol/g tissue

	R_c		R_n		TOTAL(R_c+R_n)	
	-NaSCN	+NaSCN	-NaSCN	+NaSCN	-NaSCN	+NaSCN
Expt. 1	670	890	846	1436	1517	2326
Expt. 2	429	588	973	1381	1402	1969

Table 1. Cytosolic and nuclear fractions were incubated at 0°C for 72 h with various concentrations of [^{3}H] mibolerone in absence or presence of 0.15M NaSCN. Nonspecific binding was determined in samples incubated with 1000 fold excess of

unlabeled DHT. Specific binding data were analyzed according to Scatchard and the maximal number of binding sites is presented.

Steroid Specificity:

In order to determine if under these exchange conditions [^{3}H]DMNT binding is specific for AR, cytosolic and nuclear fractions obtained from patients with benign prostatic hypertrophy (BPH) were incubated at 0°C with NaSCN (0.15M) and sucrose (15%) for 72 h with [^{3}H]DMNT in the absence or presence of a 100 or a 1000 fold excess of selected unlabeled steroids. As shown in Table 2, the three potent androgens DMNT, R1881, and DHT, effectively competed for receptor binding while progestin and estrogen showed slight competition only at high steroid concentrations, and cortisol did not compete at all.

SPECIFICITY OF [^{3}H]DMNT BINDING TO CYTOSOLIC AND NUCLEAR ANDROGEN RECEPTORS OF HUMAN PROSTATE

	CYTOSOLIC AR		NUCLEAR AR	
	1000X	100X	1000X	100X
DMNT	0	0%	0	0
DHT	15	9%	20	18%
R1881	3	10%	10	10%
ESTRADIOL	60	83%	67	81%
PROGESTERONE	50	86%	61	90%
CORTISOL	94	96%	100	100%

TABLE 2. Cytosols and nuclear suspensions were incubated at 0°C with 0.15M NaSCN, 15% sucrose, and 5nM [^{3}H]DMNT in absence or presence of a 100 or 1000 fold molar excess of unlabeled steroids. To determine nonspecific binding parallel incubations were made with [^{3}H]DMNT and a 1000 fold molar excess unlabeled DMNT.

Measurement of AR in Cytosolic and Nuclear Fractions of Dog Prostate.

Since dog prostatic tissue is frequently used as a model for the study of BPH it was determined if this exchange assay can also be used for measurement of AR in this species. The data in Fig. 4 show that exchange of cytosolic

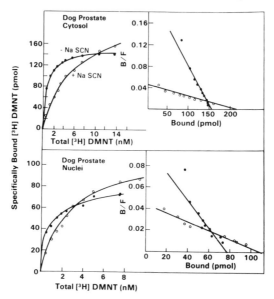

Fig. 4   Quantitation of AR in Dog Prostatic Cytosol and Nuclei. Aliquots of cytosol (5 mg protein/ml) and nuclear suspensions (137.5 ug DNA/ml) were incubated at 0°C for 72 h with an equal volume of buffer containing increasing concentrations of [^{3}H]DMNT and 15% sucrose in the absence (closed circles) or presence (open circles) of 0-15M NaSCN. Bound radioactivity was determined by HAP assay and specific binding was normalized according to Scatchard.
Panel A   Cytosol without (-) NaSCN (solid circles) (61 fmol of AR/mg protein), cytosol with (+) NaSCN (open circles) (88 fmol of AR/mg protein).
Panel B   Nuclei   1120 fmol AR/mg DNA (- NaSCN) (solid circles)
                   1600 fmol AR/mg DNA (+ NaSCN) (open circles)

and nuclear AR in the presence of NaSCN at 0°C yielded a significantly greater number of binding sites than in its absence, regardless of the length of incubation. This suggests the procedure is readily adaptable to quantitative measurement of AR in the dog prostate.

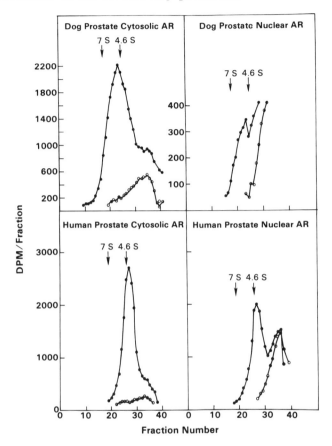

Fig. 5 SDG Analysis of Cytosolic and Nuclear AR Labeled by Exchange. Aliquots of cytosol and nuclear suspension from dog (Upper Panel) and human (Lower Panel) prostates were incubated with [^{3}H]DMNT, NaSCN (0.15M), and sucrose (15%) for 72 h at 0°C. At this point, the nuclear incubation received a small volume of buffer containing KCl to give a final concentration of 0.4M and was incubated for 45 min at 0°C. HAP slurry (0.5ml) was added to all incubations and free steroid and sucrose were removed by washing. The labeled AR was eluted from HAP with 0.4 M K_2HPO_4 buffer (pH 7.6) and aliquots were analyzed on SDG containing 0.4M KCl and 50 mM molybdate. Total binding is represented by solid circles, nonspecific binding is represented by open circles.

SDG Analysis of AR Labeled by Exchange at 0°C in the Presence of NaSCN.

AR labeled with [^{3}H]DMNT in the presence of NaSCN/sucrose and incubated for 72 h at 0°C sedimented as a 4-4.6S entity on SDGs containing 0.4M KCl and 50mM molybdate (Fig. 5). This sedimentation property is characteristic of androgen receptor complexes (Robel et al 1985).

Analysis of AR in Small Tissue Samples by NaSCN Exchange Assay.

Because the amount of tissue that can be obtained by needle biopsy is very small (50-100 mg tissue) attempts were made to scale down the procedure to measure AR in cytosolic and nuclear fractions of such small tissue samples. As shown in Table 3 cytosolic and nuclear ARs measured in 50 mg of homogeneous tissue powder (n=6) were similar to those measured in 100-400 mg tissue. Thus the sensitivity and reproducibility of this assay is not compromised by down-scaling.

AR IN HUMAN PROSTATE

(fmol/g tissue)

No. of Expt	Tissue Weight (mg)	R_c	R_n	TOTAL
6	50±4	762±238	1143±428	1906±541
6	100±4	751±212	1102±249	1853±338
6	200±7	866±185	1302±354	2179±271
3	400±19	912±252	957±45	1870±295

TABLE 3. Prostate tissues from several patients were pooled and pulverized; aliquots of the tissue powder were weighed out and kept frozen at -80°C. Analysis of AR in these tissue samples was carried out by homogenization in 10 volumes of buffer and the subcellular fractions were incubated (triplicates) at 0°C for 72 h with 0.15M NaSCN and 20nM [^{3}H] mibolerone. Nonspecific binding was determined as described in

Methods; only specific binding is shown.

DISCUSSION

Several attempts have been made to correlate the concentration of androgen receptors in human prostatic cancer tissue with clinical response or duration to hormone therapy (Connolly and Mobbs 1984), (Ekman et al 1978), (Fentie et al 1986), (Ghanadian et al 1981), (Gonor et al 1984), (Trachtenberg and Walsh 1982), (Wagner 1978). A wide spectrum of irreconcilable results have been obtained varying from no correlation (Wagner 1978) to correlation with total AR (Ghanadian et al 1981), to significant correlation only with nuclear AR (Fentie et al 1986), (Gonor et al 1984), (Trachtenberg and Walsh 1982). These discrepancies may be due in part to deficiencies in the methods applied to the measurement of AR. The work described here was undertaken to validate measurements of AR by an exchange assay in human and dog prostate tissue fractions by a procedure recently developed for evaluation of rat ventral prostatic AR (Traish et al 1985b), as well as to scale down the assay for use with tissue samples comparable in size to needle biopsies. The exchange procedure takes advantage of the stabilization of AR by sucrose and the increase in the rate of exchange induced by NaSCN at 0°C (Traish et al 1985b).

From the data presented it is clear that for both dog and human prostate a greater number of androgen binding sites was measured when exchange was carried out at 0°C for 72 h in the presence of NaSCN, as compared with that obtained by incubation at 0°C for 24-72 h. This new assay overcomes the two prinicpal difficulties normally encountered in measurements of AR by exchange, namely slow dissociation of endogenously bound ligand from AR at 0°C, and inactivation of AR at elevated temperature. We have shown that the exchanged receptor has a high affinity for [^{3}H]DMNT (K_D=2-5nM), that it sediments as a 4-4.6S moiety on KCl-containing gradients, and that the binding is specific for androgens since estradiol, progesterone and cortisol did not effectively compete for [^{3}H]DMNT binding.

The data presented in Table 3 clearly show that this assay is applicable to quantitation of AR in small tissue samples approximating those obtained from needle biopsy specimens without compromising the sensitivity of the assay.

The applicability of this assay to measurement of AR in tissue samples of as little as 50 mg should make it possible to address the question of whether or not AR levels in prostatic tissue can serve as a therapeutic guide and prognostic factor in the management of prostatic cancer patients.

The use of [^{3}H]DMNT as a ligand offers several advantages such as low nonspecific binding, absence of specific binding to sex steroid binding globulin, and photostability. Although [^{3}H]DMNT binds to progesterone receptors in human tissues (Schilling and Liao 1984), (Traish et al 1986a), (Traish et al 1986b) use of unlabeled DHT to compete with [^{3}H]DMNT binding allows determination of nonspecific binding and measurement of androgen binding sites only, even in the presence of progesterone receptors.

Since this assay procedure was found to be equally useful in the measurement of AR in human, rat, and dog prostate, and in dog testis (data not shown) it is likely that it could be applied to other tissues also.

In summary, we have validated an exchange assay for measurement of AR binding sites in human and dog prostatic tissue fractions at 0°C. The method provides significantly improved recoveries of AR compared to procedures previously described.

REFERENCES

Barrack ER, Bujnovszky P, Walsh PC (1983). Characterization of nuclear salt resistant receptors. Cancer Res 43:1107.
Carroll SL, Rowley DR, Chang CH, Tindall DJ (1984). Exchange assay for androgen receptors in the presence of molybdate. J Steroid Biochemistry 12:353.
Connolly JG, Mobbs BG (1984). Clinical applications and values of receptor levels in treatment of prostate cancer. The Prostate 5:477.
Ekman P, Snochowski M, Zetterberg A, Hogberg B, Gustafsson JA (1978). Steroid receptor content in human prostatic carcinoma and response to endocrine therapy. Cancer 44:1173.
Fentie DD, Lakey WH, McBlain WA (1986). Applicability of nuclear androgen receptor quantification to human prostatic adenocarcinoma. J Urol 135:167.
Fiet EI, Muldoon TG (1983). Differences in androgen binding

properties of the two molecular forms of androgen receptors in rat ventral prostate cytosol. Endocrinology 112:592.

Ghanadian R, Auf G, Williams G, Davis A, Richards A (1981). Predicting response of prostatic carcinoma to endocrine therapy. The Lancet II:1418.

Gonor SE, Lakey WH, McBlain WA (1984). Relationship between concentrations of extractable and matrix-bound nuclear androgen receptor and clinical response to endocrine therapy for prostatic adenocarcinoma. J Urol 13:806.

Hechter O, Mechaber D, Zwick A, Campfield LA, Eychenne B, Baulieu E-E, Robel P (1983). Optimal radioligand exchange conditions for measurement of occupied androgen receptor sites in rat ventral prostate. Arch Biochem 224:49.

Hicks LL, Walsh PC (1979). A microassay for the measurement of androgen receptor in human prostatic tissue. Steroids 33:389.

Olsson CA, White RD, Goldstein I, Traish AM, Muller RE, Wotiz HH (1979). A preliminary report on the measurement of cytosolic and nuclear prostatic tissue steroid receptors. In "Progress in Clinical and Biological Research," Murphy GP, Sandberg AA (eds) New York: Alan R. Liss, p 209.

Robel P, Eychenne B, Blondeau J-P, Baulieu E-E, Hechter O (1985). Sex steroid receptors in normal and hyperplastic human prostate. The Prostate 6:255.

Schilling K, Liao S (1984). The use of radioactive 7α, 17α-dimethyl-19-nortestosterone (mibolerone) in the assay of androgen receptors. The Prostate 5:581.

Shain SA, Boesel RW, Lamm DL, Radwin HH (1978). Characterization of unoccupied (R) and occupied (RA) androgen binding components of the hyperplastic human prostate. Steroids 31:541.

Shain SA, Gorelic LS, Boesel RW, Radwin HM, Lamm DL (1982). Human prostate androgen receptor quantitation: Effects of temperature on assay parameters. Cancer Res 42:4849.

Snochowski M, Pousette A, Ekman P, Bression D, Andersson L, Hogberg B, Gustafsson JA (1977). Characterization and measurement of androgen receptor in human benign prostatic hyperplasia and prostate carcinoma. J Clin Endocrinol Metab 45:920.

Trachtenberg J, Walsh PC (1982). Correlation of prostatic nuclear androgen receptor content with duration of response and survival following hormonal therapy in advanced prostate cancer. J Urol 127:466.

Traish AM, Muller RE, Wotiz HH (1984). Differences in the

physiocochemical characteristics of androgen receptor complexes formed in vivo and in vitro. Endocrinology 114:1761.
Traish AM, Muller RE, Wotiz HH (1985a). Resolution of non-activated and activated androgen receptors based on differences in their hydrodynamic properties. J Steroid Biochem 22:601. 1397.
Traish AM, Muller RE, Wotiz HH (1985b). A new exchange procedure for the quantitation of prostatic androgen receptor complexes formed in vivo. J Steroid Biochem 23:405.
Traish AM, Muller RE, Wotiz HH (1986a). Binding of 7α, 17α-dimethyl-19-nortestosterone (mibolerone) to androgen and progesterone receptors in human and animal tissues. Endocrinology 118:1327.
Traish AM, Williams DF, Wotiz HH (1986b). Binding of [^{3}H] 7α, 17α-dimethyl-19-nortestosterone (mibolerone) to progesterone receptors: comparison with binding of [^{3}H] R5020 and [^{3}H] ORG2058. Steroids 47:157.
Wagner RK (1978). Extracellular and intracellular steroid binding proteins. Properties, discrimination assays, and clinical applications. Acta Endocr 88:5 (Suppl 218).
Wilson EM, French FS (1976). Binding properties of androgen receptors. J Biol Chem 251:5620.

PROGESTIN EFFECTS ON LACTATE DEHYDROGENASE AND GROWTH IN THE HUMAN BREAST CANCER CELL LINE T47D

Michael R. Moore, Rodney D. Hagley and Judith R. Hissom
Department of Biochemistry
Marshall University School of Medicine
Huntington, WV 25704

INTRODUCTION

Although much is known about estrogen effects on specific proteins and growth in target tissues for female steroid hormones, knowledge about the effects of progestins is comparatively limited. In the case of mammary tissue, Fournier et al. have described a progestin-responsive 17β-hydroxysteroid dehydrogenase in surgically removed human mammary fibroadenomas (Fournier et al. 1982). In patients given exogenous progestins and during the luteal phase of the menstrual cycle, enzyme activity increased. Mauvais-Jarvis and his co-workers (Prudhomme et al., 1984; Gompel et al, 1986) have also reported the progestin stimulation of 17β-hydroxysteroid dehydrogenase in normal human breast epithelial primary cultures and in progesterone-receptor positive human breast tumors (Fournier et al., 1985). Most information on mammary gland-associated progestin-responsive proteins, however, have been obtained with breast cancer cell lines in long term tissue culture, especially using the cell line T47D, originally isolated from a pleural effusion of mammary adenocarcinoma (Keydar et al., 1979). Found by Horwitz et al. to have high levels of progesterone receptor (Horwitz et al., 1978b), this line has proven particularly useful in studying progestin effects. Judge and Chatterton have reported an increase in triglycerides accompanied by accumulation of cytoplasmic lipid vacuoles in these cells, although a pharmacological dose (10uM) of progestin was required to produce maximal response (Judge and Chatterton, 1983). Chalbos and Rochefort have found progestins to stimulate synthesis of a Mr 48,000 protein released into the medium by T47D cells (Chalbos and Rochefort, 1984a). They used [^{35}S]methionine labelling followed by electrophoresis and autoradiography in this report as in a later one in which they (Chalbos and Rochefort, 1984b) described a Mr 250,000 intracellular protein stimulated by progestins in both T47D and MCF-7 cells. Also using the MCF-7 line, Butler and his colleagues have described progestin-responsive plasminogen activator activity, but only at supraphysiological progestin concentration (10^{-6} M) except when estradiol was also present (Butler et al., 1985). Sutherland and co-workers (Murphy et al., 1986a) have found that lactogenic receptors (for growth hormone and prolactin) are stimulated by progestins in both T47D and MCF-7 cells and that epidermal growth factor receptors are progestin-elevated in T47D cells (Murphy et al., 1986b). Horwitz and Freidenberg have reported progestin stimulation of insulin receptors in T47D$_{co}$, a variant subline of T47D cells (Horwitz and Fridenberg, 1985). We recently reported (Hagley and Moore, 1985) the progestin stimulation of lactate dehydrogenase (LDH) in T47D cells obtained from the American Type Culture Collection. In this article we will describe our further characterization of this phenomenon. We will also outline how the recent discovery that phenol red, the virtually universally used tissue culture medium pH indicator, is an estrogen (Berthois et al., 1986) led us to the finding that progestins stimulate rather than inhibit the growth of T47D cells. We will also mention possible therapeutic implications of this phenomenon.

MATERIALS AND METHODS:

Cell Culture

T47D cells, supplied by the American Type Culture Collection, were grown in plastic tissue culture flasks (75cm^2) in 5% CO_2 in air at 37°C. Growth medium consisted of basal medium Eagle: modified-autoclavable (powdered) with added nonessential amino acids, 2mM L-glutamine, 10% fetal calf serum (heat-inactivated), penicillin (100 units/ml), streptomycin (100 ug/ml) (Grand Island Biological Co.) and insulin (6 ng/ml) from Sigma. To harvest the cells, growth medium was replaced with Hank's balanced salt solution without calcium and magnesium, but including 1 mM EDTA, and cells were incubated 10 minutes at 37°, shaken off the flasks, and centrifuged. Cells were then washed with 0.9% NaCl, centrifuged, and either frozen at -70°C or homogenized immediately.

To set up an experiment, approximately 2.5×10^6 cells were plated into each T-75 flask containing the above medium except that dextran-coated charcoal-treated fetal calf serum was used (Horwitz and McGuire, 1978). After 2-3 days growth in this medium, hormones were added in absolute ethanol, control flasks receiving ethanol only, to give a final ethanol concentration of 0.1%. Cells were then grown for varying lengths of time, medium changed as indicated in the figure legends.

Extraction and Assay of Lactate Dehydrogenase

Homogenization of cell pellets was done in 2 ml of 0.9% NaCl with a motor-driven, steel-shafted, teflon-tipped pestle and a conical ground-glass homogenizer (Kontes). Homogenization consisted of 10 up-and-down strokes with 30 seconds cooling in ice between each series of 10 strokes for a total of 30 strokes. Nuclei and other particulate matter were then pelleted at 46,000 xg for 1 hr at 2°. The supernatant was assayed for lactate dehydrogenase at 340 nm using a Gilford 250 spectrophotometer and 6050 recorder with the assay mixture described by Wacker (Wacker et al., 1956). Protein concentrations were determined by the method of Lowry et al. (Lowry et al., 1951) or Bradford (Bradford, 1976). DNA was estimated by the method of Burton (Burton, 1956). Enzyme units are micromoles of NADH formed per minute.

Cell Growth Experiments

Approximately 2.5×10^6 cells were plated into each Nunc plastic flask (80cm^2) with the above medium containing 10% fetal calf serum (charcoal-stripped) but without phenol red. After 3 days growth, medium was changed and hormones were added in absolute ethanol, control flasks receiving ethanol alone. Cells were then grown for the times indicated in the figures, medium changed as indicated. DNA determinations were done on the whole cells by the method of Burton (Burton, 1956).

Electrophoresis

The Beckman Paragon LD Lactate Dehydrogenase Isoenzyme Electrophoresis kit was used for LDH isozyme electrophoresis on prepared agarose slab gels buffered at pH 8.2. After electrophoresis at 100 volts for 20 minutes, the gel was soaked in substrate solution containing lactate and NAD to form NADH. This, in the presence of phenazine methosulfate, reduces nitroblue tetrazolium chloride, giving a blue color at the position of enzyme activity. Gels were fixed with 5% acetic acid, dried for 15 minutes at 90° and scanned at 600 nm.

Chemicals

Sigma Chemical Co. was the source of progesterone, testosterone, estradiol-17β, hydrocortisone, L(+) lactic acid (grade L-1), NAD (Grade V), cycloheximide and actinomycin D. R5020 (17,21-dimethyl-19-nor-4,9-pregnadiene-3,20-dione) was purchased from New England Nuclear. RU486 (RU38486) was a gift from Dr. R. Deraedt of Roussel-Uclaf.

RESULTS AND DISCUSSION

Since the human breast cancer cell line T47D has high levels of progesterone receptor (Horwitz et al., 1978), it is a good line in which to study various aspects of progestin action. Burke et al. have shown that lactate dehydrogenase in MCF-7 cells, which have high levels of estrogen receptor, is stimulated by estradiol-17β but not by progesterone (Burke et al., 1978). Different proteins in other systems have been found to be responsive to both estrogen and progesterone, often requiring priming by estrogen, presumably to produce sufficiently high quantities of progesterone receptor. Since T47D cells have high levels of progesterone receptor, we reasoned that progestins alone at physiological levels might stimulate lactate dehydrogenase activity. This idea proved correct, as 10^{-9} M progesterone increased LDH specific activity to about 150% of the control value after 5 days of treatment, changing the medium every day. The same phenomenon occurred whether analyzed per mg protein or per mg DNA, as in all other experiments reported in this article. Treatment with the synthetic progestin R5020 at 10^{-9} M gave a similar effect, except that the medium had to be changed only every 2 days, since R5020 is stable whereas progesterone is rapidly metabolized (Horwitz et al., 1983). As mentioned earlier, Burke et al. found that in MCF-7 cells, physiological levels of estradiol-17β, but not progesterone, stimulated lactate dehydrogenase (Burke et al., 1978). This may be related to the fact that MCF-7 cells contain high levels of estrogen receptor but lower levels of progesterone receptor than T47D cells (Horwitz et al., 1978).

We next determined the concentration dependence of progestin stimulation of LDH, finding that stimulation increases from 10^{-11} M to 10^{-9} M, where it plateaus, through 10^{-7} M. Thus, the effect occurs in the physiological range of progestin concentration (Fournier et al., 1982) in a dose-responsive manner.

We next determined the time dependency of the phenomenon as shown in figure 1. Stimulation plateaus after 2-3 days treatment, is maintained through day 4, and then falls off with time. Although there is no day one time point shown in this particular experiment, other experiments have shown the time dependency from day zero to day 2 to be as the line shows. Similar patterns of elevation of specific proteins, followed by a decline, have been observed by other authors in T47D cells in response to R5020 treatment. Chalbos and Rochefort have described progestin induction of a Mr 48,000 protein which reaches a maximum at 3-4 days and then declines (Chalbos and Rochefort, 1984). R5020 stimulation of insulin receptor as reported by Horwitz and Freidenberg follows a similar pattern (Horwitz and Freidenberg, 1985). Noting that release of T47D cells from growth rate inhibitory effects of R5020 occurs at about the same time as the decline of insulin receptor levels, Horwitz and Freidenberg have speculated that the two may be associated. Although we have also observed a slight R5020 inhibition of growth of T47D cells (not statistically significant), we are uncertain as to the significance of the rise-decline pattern of LDH stimulation.

In order to ascertain which LDH isozymes were present in T47D cells, we performed isozyme electrophoresis on cell extracts, using the Beckman Paragon Electrophoresis System. We found only one LDH isozyme, LDH-5, made up of 4 type M subunits. This is the isozyme found in greatest abundance in skeletal muscle. Upon treatment with R5020, LDH-5 re-

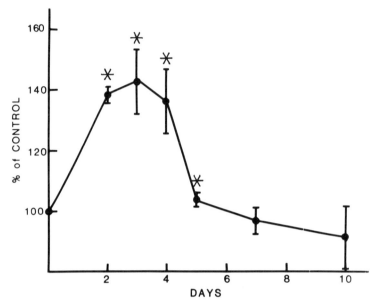

Figure 1: Time dependency of LDH stimulation by R5020. After plating and maintenance in charcoal-stripped serum-containing medium for 2 days, cells were treated with 10^{-9} M R5020 for various time periods, medium changed every 48 hours. For every time point there were triplicate progestin-treated and control flasks, which received ethanol alone. Activity is in terms of units/mg protein and data points refer to mean $\pm$ standard deviation for triplicate flasks. This experiment was done 3 times with similar results. *, statistically different from control at p<0.05 by Student's unpaired t test.

---

mained the only isozyme. These results are similar to those of Burke et al., who found only the LDH-5 isozyme in MCF-7 cells, both with and without estrogen (Burke et al., 1978).

The LDH-5 isozyme is particularly well suited for a role in anaerobic metabolism as it has an especially low Km for pyruvate and high Vmax for lactate formation. Since varying proportions of hypoxic cells exist in a tumor mass, LDH-5 levels may be important for their survival. In fact a link between LDH and survival of malignant cells has been postulated since Warburg in 1930 showed they contained high levels of lactate (Warburg, 1930). Other workers (Goldman et al., 1964) have found that LDH is elevated in neoplastic disease, Hilf et al. showing a doubling of LDH activity in breast cancer tissue as compared to benign proliferative disease and normal breast (Hilf et al., 1976). Perhaps progestin stimulation of LDH-5 in vivo helps breast cancer cells survive in a comparatively hypoxic environment.

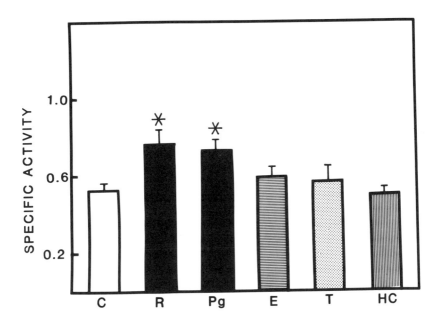

Figure 2: Specificity of steroid hormone effects on lactate dehydrogenase. After seeding and growth for 3 days in charcoal-stripped serum-containing medium, cells were treated with ethanol alone or 10^{-8} M hormone in ethanol for 3 days, medium changed every day. Data are in terms of units/mg protein and refer to mean $\pm$ standard deviation for triplicate flasks. E: estradiol-17β; R: R5020; Pg: progesterone; T: testosterone; HC: hydrocortisone; C: control. This experiment was done 3 times with similar results. *, statistically different from all others (except each other) at p<0.05 by Student-Newman-Keul's multiple comparison procedure. No other differences are significant.

As shown in figure 2, the effect is specific for progestins. We did find a stimulation by estradiol-17β when 2 particular lots of serum were used; this stimulation, however, was elusive and not consistently reproducible. The effects of progestins are very reproducible and have occurred with all serum lots we have used (eight lots). We think the stimulation of LDH by progestins probably occurs through the progesterone receptor for three main reasons. First of all, as shown in figure 2, the effect is specific for progestins. Secondly, we have shown the effect in the progestin receptor-rich line T47D, whereas Burke et al. did not see it in MCF-7 cells, which have lower levels of progestin receptor. Thirdly, as shown below, the antiprogestin RU486 (17 β -hydroxy-11β -(4-dimethylaminophenyl-1)-17 α (prop-1-ynil)-estra-4,9-dien-3-one) prevents progestin stimulation of LDH.

In order to determine whether the synthesis of new RNA is required, we utilized the DNA intercalator actinomycin D at 1 ug/ml. Previously, we determined that RNA synthesis was inhibited to less than 5 percent of control and that cell viability was unaffected under the conditions of these experiments. The data of figure 3 suggest that new RNA synthesis is required for the effect of progestins on lactate dehydrogenase. We don't know at this stage whether the RNA is message for LDH or some other RNA necessary for the induction.

Our next experiment was designed to gain information about the role of protein synthesis in LDH stimulation by progestins. After seeding cells in charcoal-stripped serum-containing medium and growth for 2 days, control cells and progestin-treated cells were simply medium changed. Cycloheximide at a final concentration of 1mM was then added to the appropriate cells after 24 hours of treatment with control or 10^{-8} M progestin-containing medium. After 4 hours of cycloheximide treatment, the medium was changed to fresh medium with or without progestin and cells were grown for 20 more hours, when the cells were harvested and assayed for LDH. Cycloheximide treatment had no effect on viability of the cells. Results of 3 separate experiments showed that cycloheximide treatment prevented the stimulation of LDH by R5020, just as in the case of actinomycin D. Although these data suggest that new protein synthesis is probably required for the LDH stimulation, we do not know at this point whether

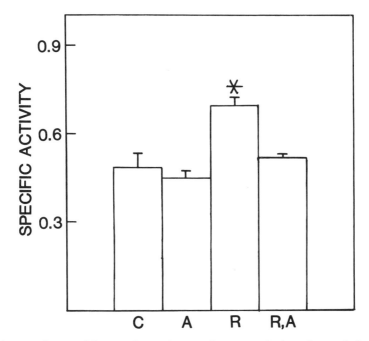

Figure 3: Effect of actinomycin D on induction of lactate dehydrogenase. Two days after plating in charcoal-stripped serum-containing medium, actinomycin D treatment was begun in the indicated cells for 1 hour. After this, medium was changed in all flasks with addition of either nothing (control), 10^{-8} M R5020, 1 ug/ml actinomycin D, or actinomycin D plus 10^{-8} M R5020. Actinomycin D was dissolved in H_2O and R5020 in ethanol. Control flasks received solvents only. After treatment for 24 hours, cells were harvested, extracted, and assayed for LDH. Data refer to mean $\pm$ standard deviation for triplicate flasks and specific activity is in units per mg protein. This experiment was done three times with similar results. *, statistically different from all others at $p<0.01$ by Student-Newman-Keul's multiple comparison procedure. No other differences are significant.

---

it is new LDH or some other protein(s) or both. Stimulation of enzyme activity is not dependent on DNA synthesis, as 24 hr treatment with 10ug/ml ara C (cytosine-β-D-arabinofuranoside) still allows progestin stimulation of LDH.

If the effect of progestins on LDH is occurring through the progesterone receptor, it should be inhibited by the progestin antagonist RU486 (Herrman et al., 1983). As shown in figure 4, RU486 does prevent the stimulation. RU486 is also an antiglucocorticoid (Jung-Testas and Baulieu, 1983), but the lack of stimulation of LDH by hydrocortisone is again consistent with the idea that stimulation occurs through the progesterone receptor.

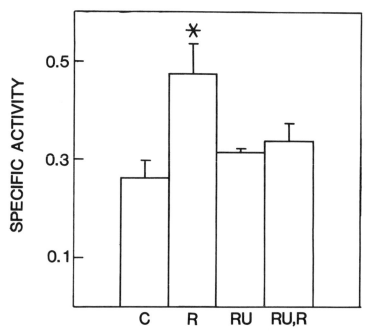

Figure 4: Effect of the anti-progestin RU486. Cells were seeded as in figure 3 and grown for 2 days. At this time, medium was changed to fresh with either 10^{-9}M R5020 (R), 10^{-8}M RU486(RU), R5020 and RU486 (RU + R), or with solvent alone (C). Solvent was 100% ethanol, which was at a final concentration of 0.2% in all flasks. Data refer to mean ± standard deviation of triplicate flasks. This experiment was done 3 times with similar results. *, statistically different from all others at p<0.01 by Student-Newman-Keul's multiple comparison procedure. No other differences are significant.

After we had completed these experiments, Berthois et

al. reported the finding that phenol red, the pH indicator used in virtually all tissue culture media, has estrogenic activity at the concentrations used in tissue culture (Berthois et al., 1986). The authors showed that phenol red binds to the estrogen receptor, stimulates the growth of the estrogen-receptor rich human breast cancer cell line, MCF-7, and stimulates the level of progesterone receptor. Since the presence of phenol red was probably affecting the above results, we tested for R5020 stimulation of LDH in cells grown in the absence of phenol red. As shown in figure 5, we still observed stimulation.

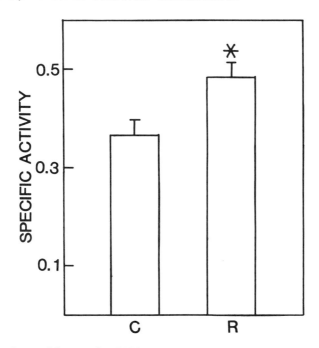

Figure 5. Effect of R5020 on LDH in the absence of phenol red. This experiment was done as described for R5020 and control in figure 2 except that there was no phenol red in the medium, either during the 3 days after plating, prior to addition of hormone, or during hormone incubation. The difference is significant at $p<0.01$ by Student's unpaired t test.

While repeating the experiment in figure 5, we noticed that the R5020-treated cells seemed to grow more quickly than the control cells when there was no phenol red. It had previously been shown by Horwitz and co-workers that R5020 inhibits the growth of T47D$_{co}$ cells, an estrogen insensitive subline of T47D cells (Horwitz and Freidenberg, 1985). Vignon et al. had found that progestins had no effect alone but inhibited the stimulatory effects of estradiol on growth of the clone 11 subline of T47D cells (Vignon et al., 1983). Gompel et al. had also reported R5020 inhibition of growth of primary cultured normal human breast cells (Gompel et al., 1986). In our hands, R5020 had had no effect on growth of the T47D cells we obtained from the American Type Culture Collection, although sometimes there was a slight inhibition. All of the above data on growth effects of progestins had been obtained in the presence of phenol red, before it was known that phenol red was an estrogen.

Noticing the seeming stimulation of growth in our LDH stimulation experiments in the absence of phenol red, we decided to pursue growth studies further.

Figure 6 shows the effects of R5020 on T47D cell growth in the presence and absence of phenol red. As shown, when the cells were grown in phenol red-containing medium, 10^{-8}M progestin treatment for 10 days had no statistical effect on cell growth, although the progestin-treated cells appeared to grow slightly slower than control cells. In the absence of phenol red, however, 10^{-8}M progestin-treated cells grew faster over the 10 day period, so that there was 2 x as much DNA in the progestin-treated flasks as in control flasks. This stimulation of growth by progestin occurred not only at the 10 day time point, but throughout the entire 10 day growth period (data not shown).

We next reasoned that if progestins stimulated growth to this extent in the absence of phenol red, perhaps estrogens would stimulate even more, since Rochefort and his co-workers (Chalbos et al., 1982) had already shown an estrogen stimulation of T47D clone 11 even in the presence of phenol red. Figure 7 shows that estradiol-17β (10^{-8}M) stimulated growth 6 fold over control in 8 days of hormone treatment. Other experiments have shown an even greater stimulation. R5020 again stimulated growth, and a combination of estradiol and R5020 showed the antiestrogenic effect of the progestin on growth.

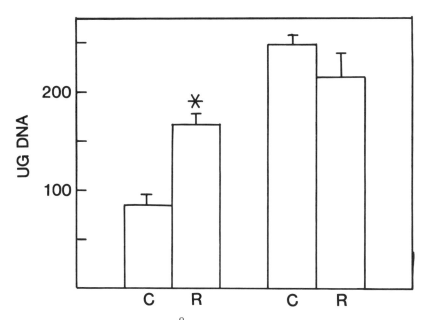

Figure 6: Effect of 10^{-8}M R5020 on cell growth in the presence and absence of phenol red. Cells were plated as in figure 5 either in the absence or presence of phenol red. After 10 days growth, medium changed every 2 days, cells were harvested and assayed for DNA in the growth flasks. Data refer to mean ± standard deviation for triplicate flasks in the case of no phenol red (representative of 4 experiments). The data from cells grown in the presence of phenol red are the average ± standard deviation of duplicate flasks (representative of 3 experiments). *, different from control at p<0.01 by Student's unpaired t test.

These data are very interesting to us because they have unmasked progestin stimulation of human breast cancer cell growth in tissue culture. They also show very clearly the antiestrogenic effects of the progestin. These stimulatory effects of the hormones again were evident not only at the end of the experiment, but throughout the 8 days of growth. The control cells, however, actually began to die off during the 8 days growth and, in some experiments, there were actually less cells at the end of 8

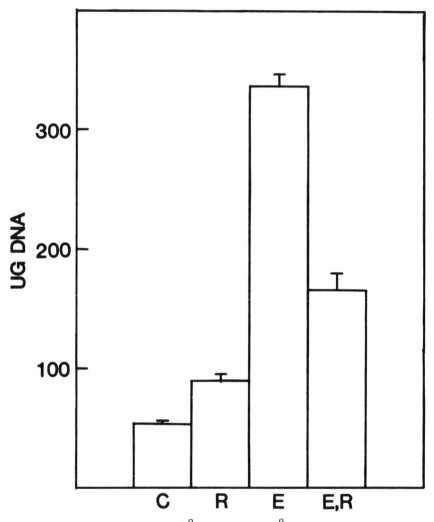

Figure 7: Effects of 10^{-8}M R5020, 10^{-8}M estradiol-17β, and a combination of the two on cell growth in the absence of phenol red. Cells were plated as described in figure 6 and grown for 8 days in the presence or absence of hormones, medium changed every 2 days. Final ethanol concentration was 0.2% in all flasks. Data refer to mean ± standard deviation for triplicate flasks. All differences are statistically significant at p<0.05 by Student-Newman-Keul's multiple comparison procedure.

days than at the start. It began to look as if the T47D cells might actually require estrogen or progestin for continued growth; i.e., it began to look as if the cells might be dependent on, rather than simply responsive to, these hormones.

Our next experiment was to determine whether the antiprogestin RU486 would prevent R5020 stimulation of growth. As shown in figure 8, the antiprogestin alone had no effect on growth, but completely prevented the stimulation of growth by R5020.

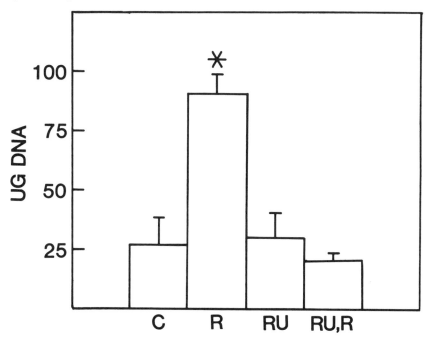

Figure 8: Inhibition by the antiprogestin RU486 of R5020 stimulation of growth. Cells were plated and grown for 8 days as described in figure 7, in the absence of phenol red. Hormone concentrations were 10^{-9}M R5020, 10^{-8}M RU486. Data refer to mean ± standard deviation of triplicate flasks. *, different from all others at $p<0.01$ by Student-Newman-Keul's multiple comparison procedure. No other differences are significant.

The above data, taken together, point to the notion

that both estrogens alone and progestins alone stimulate growth of these T47D cells. This may also be the situation in breast cancer patients with both progesterone and estrogen receptors. The data then raise the attractive possibility that a combination of antiestrogen and antiprogestin therapy might be more effective in patients than are antiestrogens alone. The response rate to tamoxifen is usually about 60-65% in patients with ER-positive tumors (Manni et al., 1980). Perhaps some of the cells in these patients' tumors are being maintained by progestins. We are aware that tamoxifen would probably lower the amount of progesterone receptors, but may leave enough for a positive growth response to progesterone, just as occurs in our T47D cells in the absence of estrogen.

Thus, the above system seems to be an attractive one for studying the effects of progestins and estrogens on breast cancer cell growth in vitro. We are currently using this model to further study the effects of combination antiprogestin and antiestrogen treatment on cell growth, in the hope of predicting more effective hormone therapy for breast cancer patients.

ACKNOWLEDGMENTS

This work was supported in part by grants CA29501 and RR05870 from the N.I.H. and by a grant from the Huntington Clinical Foundation. We thank Dr. Howard Quittner of the Marshall University Department of Pathology for use of the LDH electrophoresis system, Dr. R. Deraedt of Roussel-Uclaf for supplying us with the RU486, and Mrs. Dolores Brumfield for typing the manuscript.

REFERENCES

Berthois, Y, Katzenellenbogen, JA, Katzenellenbogen, BS (1986) Phenol red in tissue culture media is a weak estrogen: implications concerning the study of estrogen-responsive cells in culture. Proc Natl Acad Sci USA 83:2496-2500.

Bradford, MM (1976) A rapid and sensitive method for the quantitation of microgram quantities of protein utilizing the principle of protein-dye binding. Anal Biochem 72:248-254.

Burke, RE, Harris, SC, and McGuire, WL (1978) Lactate dehydrogenase in estrogen-responsive human breast cancer cells. Cancer Res 38:2773-2776.

Burton, DA (1956) A study of the conditions and mechanisms of the diphenylamine reaction for the colorimetric estimation of deoxyribonucleic acid. Biochem J 62:315-323.

Butler, WB, Kirkland, WL, Gargala, TL, Goran, N, Kelsey, WH, and Berlinski, PJ (1983) Steroid stimulation of plasminogen activator production in a human breast cancer cell line (MCF-7). Cancer Res 43:1637-1641.

Chalbos, D, and Rochefort, H (1984) A 250-kilodalton cellular protein is induced by progestins in two human breast cancer cell lines, MCF-7 and T47D. Biochem Biophys Res Commun 121:421-427.

Chalbos, D, and Rochefort, H (1984) Dual effects of the progestin R5020 on proteins released by the T47D human breast cancer cells. J Biol Chem. 259:1231-1238.

Chalbos, D, Vignon, F, Keydar, I, Rochefort, H (1982) Estrogens stimulate cell proliferation and induce secretory proteins in a human breast cancer cell line (T47D). J Clin Endocrinol Metab 55:276-283.

Fournier, S, Brihmat, F, Durand, JC, Sterkers, N, Martin, PM, Kuttenn, F, and Mauvais-Jarvis, P (1985) Estradiol 17 β-hydroxysteroid dehydrogenase, a marker of breast cancer hormone dependency. Cancer Res 45:2895-2899.

Fournier, S, Kuttenn, F, deCicco, F, Baudot, N, Malet, C, Mauvais-Jarvis, P (1982) Estradiol 17β-hydroxysteroid dehydrogenase activity in human breast fibroadenomas. J Clin Endocrinol Metab 55:428-433.

Goldman, RD, Kaplan, NO, Hall, TC (1964) Lactic dehydrogenase in human neoplastic tissues. Cancer Res. 24:389-399.

Gompel, A, Malet, C, Spritzer, P, Lalardrie, J-P, Kuttenn, F, Mauvais-Jarvis, P (1986) Progestin effect on cell proliferation and 17 β-hydroxysteroid dehydrogenase activity in normal human breast cells in culture. J Clin Endocrinol Metab 63:1174-1180.

Hagley, RD, Moore, MR (1985) A progestin effect on lactate dehydrogenase in the human breast cancer cell line T-47D. Biochim Biophys Res Commun. 128:520-524.

Herrmann, W, Wyss, R, Riondel, A, Philibert, D, Teutsch, G, Sakiz, E, Baulieu, EE (1983) The effects of an antiprogesterone steroid in women: interruption of the menstrual cycle and of early pregnancy. CR Acad Sci (D) (Paris) 294:933-938.

Hilf, R, Rector, WD, and Orlando, RA (1976) Multiple molecular forms of lactate dehydrogenase and glucose-6-

phosphate dehydrogenase in normal and abnormal human breast tissues. Cancer 37:1825-1830.

Horwitz, KB, Freidenberg, GR (1985) Growth inhibition and increase of insulin receptors in antiestrogen-resistant T47D$_{co}$ human breast cancer cells by progestins: implications for endocrine therapies. Cancer Res 45:167-173.

Horwitz, KB, McGuire, WL (1978) Estrogen control of progesterone receptor in human breast cancer. J Biol Chem 253: 2223-2228.

Horwitz, KB, Mockus, MB, Lessey, BA (1982) Variant T47D human breast cancer cells with high progesterone-receptor levels despite estrogen and antiestrogen resistance. Cell 28:633-642.

Horwitz, KB, Mockus, MB, Pike, AW, Fennessey, PV, and Sheridan, RL (1983) Progesterone receptor replenishment in T47D human breast cancer cells, roles of protein synthesis and hormone metabolism. J Biol Chem 258:7603-7610.

Horwitz, KB, Zava, DT, Thilagar, AK, Jensen, EM, McGuire, WL (1978) Steroid Receptor Analyses of Nine Human Breast Cancer Cell Lines. Cancer Res 38:2434-2437.

Judge, SM, and Chatterton, RT, Jr. (1983) Progesterone specific stimulation of triglyceride biosynthesis in a breast cancer cell line (T-47D). Cancer Res 43:4407-4412.

Jung-Testas, I, Baulieu, EE (1983) Inhibition of glucocorticosteroid action in cultured L-929 mouse fibroblasts by RU486, a new antiglucocorticosteroid of high affinity for the glucocorticosteroid receptor. Exp Cell Res 147:177.

Keydar, I, Chen, L, Karby, S, Weiss, FR, Delarea, J, Radu, M, Chaiteik, S, and Brenner, HJ (1979) Establishment and characterization of a cell line of human breast carcinoma origin. Eur J Cancer 15:659-670.

Lowry, OH, Rosebrough, NJ, Fair, AL, Randall, RJ (1951) Protein measurements with the Folin phenol reagent. J Biol Chem 193:265-275.

Manni, A, Arafah, B, and Pearson, OH (1980) Estrogen and progesterone receptors in the prediction of response of breast cancer to endocrine therapy. Cancer (Phila.) 46 (Suppl.):2837-2841.

Murphy, LJ, Murphy, LC, Stead, B, Sutherland, RL, and Lazarus, L (1986) Modulation of lactogenic receptors by progestins in cultured human breast cancer cells. J Clin Endocrinol Metab 62:280-287.

Murphy, LJ, Sutherland, RL, Stead, B, Murphy, LC, and

Lazarus, L (1986) Progestin regulation of epidermal growth factor receptor in human mammary carcinoma cells. Cancer Res 46:728-734.

Prudhomme, JF, Malet, C, Gompel, A, Lalardrie, JP, Ochoa, C, Boue, A, Mauvais- Jarvis, P, and Kuttenn, F (1984) 17 β- hydroxysteroid dehydrogenase activity in human breast epithelial cell and fibroblast cultures. Endocrinology 114:1483-1489.

Vignon, F, Bardon, S, Chalbos, D, and Rochefort, H (1983) Antiestrogenic effect of R5020, a synthetic progestin in human breast cancer cells in culture. J Clin Endoc Metab 56:1124-1130.

Wacker, WEC, Ulmer, DP, Vallee, BL (1956) Metalloenzymes and myocardial infarction II, malic and lactic dehydrogenase activities and zinc concentrations in serum. New England J Med 255:449-456.

Warburg, O (1930) "The Metabolism of Tumors," London:Constable.

Hormones, Cell Biology, and Cancer: Perspectives
and Potentials, pages 181–196
© 1988 Alan R. Liss, Inc.

THE CELLULAR RESPONSE OF HUMAN BREAST CANCER TO ESTROGEN

G. Wilding, M.E. Lippman and R.B. Dickson
Breast Cancer Section
Medicine Branch
National Cancer Institute
National Institutes of Health

I. INTRODUCTION

Breast cancer has been known for many years to be growth regulated by estrogenic hormones and by antiestrogenic antagonists in about 1/3 of clinical cases of metastatic disease (50). The proportion of primary (nonmetastatic) breast cancers which are hormone-dependent is unknown, but almost certainly larger. Since breast cancer occurs in women who have never had functional ovaries with only 1% of the frequency of that of women with intact ovaries, estrogens must be stimulatory, at least initially, in nearly all cases of breast cancer. However, the central role of estrogen in the etiology, maintenance and treatment of breast cancer has not been completely defined. To develop strategies for preventing and treating breast cancer in the future, an extensive effort to understand the mechanisms of the cellular hormonal response is warranted.

At the cellular level, steroid hormones modulate a wide variety of physiologic functions such as metabolism, differentiation and proliferation by regulating the expression of specific genes. The mechanisms by which modulation of these physiologic functions lead to the control of cancer cell proliferation have been under intensive investigation and have led to unifying links among growth factors, their receptors and oncogene products (24). The purpose of this paper will be to summarize the data addressing the hypothesis that estrogens can directly interact with receptor containing breast cancer cells to modulate gene expression and phenotypic properties. We will review the role of the estrogen receptor in this process and the recent observations that it and other steroid receptors have amino acid sequences which are highly homologous to the erb A oncogene (44). The effects of estrogens on macromolecular synthesis will be summarized. The majority of the review will focus on the concept that polypeptide growth factors may be common mediators of growth control for both estrogen regulated and autonomous breast cancer.

II. **THE ESTROGEN RECEPTOR**

Steroids exert their influence via an intracellular protein receptor. Following association with receptor, a subsequent "activation" step permits the steroid-receptor complex to interact with DNA and chromatin proteins and alters the expression of specific genes. Early studies of the estrogen receptor proposed that the unoccupied receptor was located in the cytoplasm, and, following ligand occupation, the receptor's affinity for chromatin increased and a translocation to the nucleus occurred (69). This concept may require revision since recent immunolocalization studies using monoclonal anti-estrogen receptor antibodies (42) have shown the apparent presence of unoccupied receptors in the nucleus of MCF-7 breast cancer cells (28, 75). Both forms of the estrogen receptor, occupied and unoccupied, are now thought to reside in the nucleus. In either case, activation alters the affinity of the receptor for nuclear components and permits interaction of the hormone-receptor complex with the genome. The precise nature of the estrogen receptor - nuclear interaction is unknown. Presumably the receptor interacts with both DNA and chromosomal proteins. A nuclear "acceptor" binding protein has been isolated for the uterine estrogen receptor (59) and other steroid receptors (65). In addition, the estrogen receptor forms a complex with the nuclear matrix (5), a chromatin scaffolding structure which may be involved in regulation of transcription and replication of DNA (56, 60). Interestingly, Toft and coworkers (64) have also recently shown that receptors for estrogen and other steroids associate in vitro with a 90 kd heatshock protein. This heatshock protein also associates with the Rous sarcoma virus transforming protein pp60 v-SRC, a plasma membrane protein. The function of the 90 kDa protein in receptor function and hormone action is not yet known.

The cDNAs of the estrogen, glucocorticoid and progesterone receptors recently have been cloned (11,25,27,37,73). The human estrogen receptor messenger RNA is 6322 nucleotides in length and contains an open reading frame of 1785 nucleotides which codes for a 595 amino acid protein of 66,182 daltons. Of interest, the mRNA also contains a long, 4305 nucleotide, 3'-untranslated region of unknown significance (26). Other steroid receptors such as the glucocorticoid and progesterone receptors, have similar structures (27,37). Comparison of chicken and human estrogen receptor sequences shown 80% amino acid homology, with the putative hormone and DNA binding regions showing 94% and 100% homology, respectively (44). When these two regions of the estrogen receptor were examined by site-directed mutagenesis, one region (region E) contained all of the sequences necessary to bind estradiol with high affinity. Another region (region C) was rich in cysteine and basic amino acids, and it contained structural characteristics thought to be important for DNA binding in other eucaryotic transcriptional regulatory proteins (45). Amino acid sequence comparison of the estrogen receptor, glucocorticoid receptor, progesterone receptor and the v-erb A oncogene product of the avian erythroblastosis virus revealed extensive regions of homology between the four genes (44). The most striking region of homology was region C of the estrogen receptor, thought to be the DNA binding region. This region of the estrogen receptor was 61% homologous to the corresponding region of the glucocorticoid receptor and 53% homologous to v-erb A (23,44,45). This suggested that c-erb A, the cellular counterpart of v-erb A, belongs to a multigene family of transcriptional regulatory proteins and is derived from a common primordial ancestral regulatory gene. In fact, recent studies have shown c-erb A to be the thyroid hormone receptor (74). Isolation of the steroid receptor genes will open many new avenues of investigation into the process of steroid induced cellular changes. Understanding such control mechanisms will allow us to develop new strategies for the treatment of hormonally responsive cancers.

III. ESTROGEN EFFECTS ON MACROMOLECULES

Using cloned human breast cancer cell lines, the effects of steroids, specifically estrogen, on cell growth, gene activation and enzyme activity has been explored. Cell lines such as MCF-7, respond to physiologic concentrations of estradiol by increasing their rate of proliferation (1,49) and these results can be duplicated in defined medium devoid of serum constituents (3). On the other hand, potent antiestrogens are capable of inhibiting estrogenic stimulation in cells containing estrogen receptors. This effect can be prevented by the simultaneous administration of estrogen and reversed by subsequent estrogen administration (48).

The mitogenic response of breast cancer cells to estrogen is maintained by increased activity of a variety of enzymes involved in both scavenger and de novo DNA synthetic pathways. They include: DNA polymerase, thymidine and uridine kinases, thymidylate synthetase, aspartate transcarbamylase, carbamyl phosphate synthetase, dihydrooratase and dihydrofolate reductase (1). Additional proteins whose activities are altered by estrogens are progesterone receptor, plasminogen activator and several secreted proteins of 7, 24, 52 and 160 kd (30,36,76). Though the progesterone receptor is not directly growth modulatory, the presence of progesterone receptor appears to be coupled to growth regulation by estrogen, and the progesterone receptor content of breast tumors is used clinically as a marker for hormone therapy responsiveness. Plasminogen activator is thought to contribute to tumor progression and growth by allowing the tumor to digest and traverse basement membranes (48). In purified form, the 52 kd secreted glycoprotein has been shown to be mitogenic for breast cancer cells in vitro (76). Of interest, plasminogen activator and the 52 kd protein are both serine proteases. It is conceivable that proteases may serve additional roles such as facilitating release of mitogenic growth factors such as IGF-I from carrier proteins, processing of growth factor or protease precursors to active species (41) or interacting directly with cellular receptors (56,67). The function of the other induced proteins is not known though some of the major species can be dissociated from cellular proliferation through the use of clonal variants in tissue culture (13).

Several enzymes are regulated at the transcriptional level. These include thymidine kinase and dihydrofolate reductase. Using cDNA probes, estrogens can be shown to increase the messenger RNA levels of these enzymes (40,46). For thymidine kinase, message levels increase as a result of increased transcription rather than stabilization of the message. Transcriptional control of a 600 base pair message, pS2, by estrogen has been demonstrated by Chambon and colleagues

(36); the function of the protein product of pS2 is unknown. Of recent interest has been the demonstration that the laminin receptor protein is increased by estrogen treatment of MCF-7 cells (2). The laminin receptor is thought to mediate attachment of cells to basement membrane laminin and to promote colonization of new host tissues. Finally, several investigators have isolated a collection of cDNA clones representing mRNA species which are induced within 6 hours of estrogen stimulation of human breast cancer cells (53).

IV. Autocrine Stimulated Growth in Breast Cancer

Several pieces of evidence support the hypothesis that hormones may stimulate a growth response via locally acting intermediate factors which serve as effectors of hormone treatment. First, the rate of proliferation of MCF-7 breast cancer cells is increased by plating at higher densities (35), suggesting conditioning of the media by the cells with factors which support growth. Second, using anchorage independent growth to assays for activities capable of inducing a malignant transformed phenotype, a number of growth factor activities have been isolated from breast cancer cells. Third, some of these peptide growth factors alter the phenotype of human breast cancer cells in vitro and, in some instances, are produced by the breast cancer cells in response to estrogen treatment (19,24). The characteristics and effects of some of these activities are described below.

A. Platelet Derived Growth Factor

Human breast cancer cells produce platelet derived growth factor (PDGF). When conditioned media from human breast cancer cells is applied to quiescent fibroblasts, DNA synthesis is initiated. In addition, the human breast cancer cell lines express RNA complementary to a c-sis probe and cellular and secreted polypeptides are immunoprecipitated by specific antiserum against PDGF. Its molecular weight, heat and reducing agent stability are characteristic of authentic PDGF. Upon examination of poly A selected mRNA from MCF-7 and MDA-MB-231 cells, transcripts fo both PDGF A and B chains were observed. Estrogen treatment of the estrogen responsive MCF-7 induced mRNA encoding both A and B chain (36).

Estrogen induced production of PDGF in breast cancer cells, along with other growth factors such as TGF a, may act in a paracrine manner on fibroblasts and other surrounding tissues to stimulate cell proliferation and possibly enhance tumor growth via the release of fibroblast derived growth factors such as IGF-I. It is possible that fibroblast derived IGF-I could be one of the stromal factors which are required in vivo for full estrogen effects on epithelial proliferation.

The potential influence of c-sis in human malignancies was demonstrated by Gazit and coworkers who showed that normal human c-sis is able to induce cellular transformation (22). When a normal human c-sis clone under the control of a retroviral LTR was transfected into NIH 3T3 cells, high titers of transforming activity were observed and the transformants expressed human c-sis translational products.

B. Insulin-like Growth Factors

The insulin-like growth factors (IGF-I and II) are found in the blood and mediate the anabolic effects of human growth hormone. They are polypeptides with molecular weights of 7600 and 7500 daltons, respectively, that have considerable homology to proinsulin, though they appear to interact with specific cell surface receptors. Authentic serum derived IGF-I (somato- median C) stimulates the proliferation of four human breast cancer cell lines, MCF-7, MDA-MB-231, ZR-75-1 and Hs578T. In addition, each of these lines produce and secrete IGF-I. The two highly tumorigenic estrogen independent cell lines, MDA-MB-231 and Hs578T, produce 2-10 fold more IGF-I activity than the estrogen responsive and less tumorigenic cell lines MCF-7 and ZR-75-1 (19). Using acid gel exclusion chromatography and an radioimmunoassay, Huff (29) showed chromatographic migration for breast cancer cell line derived IGF-I identical to authentic IGF-I. A complex series of mRNA species were also detected with Northern blot analysis using a cDNA probe to authentic IGF-I (38). One of these, a 300 bp mRNA corresponds to the smallest of 3 mRNA transcripts observed with poly A selected mRNA from human liver (19). Under strignent estrogen depleted conditions, studies have shown a 3-6 fold IGF-I induction with estradiol, TGF α, EGF or insulin treatment (31,32). IGF-I secretion is inhibited by growth inhibitory antiestrogens and glucocorticoids. The production and regulation of IGF-I by normal breast epithelium has not yet been examined. It remains to be seen whether IGF-I produced by breast cancer acts primarly on breast cancer cells in an autocrine stimulatory mode or on surrounding stroma to promote chemotaxis and growth.

C. Epidermal Growth Factor and Transforming Growth Factor a

Epidermal growth factor, EGF, is a progression factor of 6045 molecular weight shown to stimulate a wide variety of cells to proliferate, differentiate or both (68). It shares 40-45% homology with another progression factor termed transforming growth factor α, TGF α, which itself, is produced by transformed cells. Both interact with a common receptor located on the cell surface. The EGF receptor is an integral membrane protein which serves to transmit signals via an intrinsic EGF stimulated tyrosine kinase. Receptor clustering may also be involved in signal transduction. Autoregulation of the receptor may occur by phosphorylation of specific regulatory sites, receptor degradation or by intracellular sequestration of the receptor. Of interest, the EGF receptor is the proto-oncogene to the v-erb B oncogene of the avian erythroblastosis virus.

The role of EGF and its receptor in carcinogenesis is complex. Clearly, the addition of EGF to many cell types elicits responses associated with neoplastic transformation, such as proliferation, partial loss of density dependent inhibition of growth and dependency on serum for growth, loss of fibronectin and enhanced secretion of plasminogen activator (68). For these reasons, EGF and TGF α serve as prime candidates for autocrine control factors.

Secretion of TGF α by a large variety of tumor cells suggests involvement of TGF α in cell transformation and provides evidence supporting an autocrine stimulatory model. More direct evidence was provided by Rosenthal and coworkers who transfected Fischer rat fibroblasts with human TGF α cDNA construct. Synthesis and secretion of TGF α by these cells resulted in the anchorage independent growth and induced tumor formation in nude mice. In addition, anti-human TGF α antibodies prevented the expressing cells from forming colonies in soft agar (61).

Human breast cancer cell lines show a proliferative response to EGF and TGF α (17). In fact, they also secrete material of the EGF/TGF α class (62) as assessed by stimulation of anchorage-independent growth of rodent NRK and AKR-2B fibroblasts in soft agar culture. As shown by Bates et al (6), hormone-dependent and independent breast cancer cells produce a major species of approximately 30 kd of transforming activity for NRK fibroblasts. Estrogen induces this activity in estrogen-responsive cell lines, whereas it is secreted constitutively by hormone-independent breast cancer cell lines such as MDA-MB-231 cells. This species also coincides with a peak of MCF-7 antostimulatory activity and the principle species of EGF receptor competing activity (6). Antisera specific to TGF α reacts with this species (58) but it is significantly larger than the sequenced 6 kd species from transformed rodent fibroblasts (15). It is not yet certain whether this protein is related to the 17-19 kd TGF α precursor protein observed in fibroblasts (33), it is modified by glycosylation or palmitoylation (8), it is the product of alternative splicing or if it is a novel TGF α related gene. The expected 4.8 kd TGF α mRNA species has been detected in MCF-7 cells and other human breast cancer cell lines and tumors. No correlation of TGF α mRNA expression was observed with estrogen receptor status; 70% of the adenocarcinomas contained TGF α mRNA. When MCF-7 cells were treated with estradiol in vitro, TGF α mRNA was induced in 6 hrs. (7). Estrogen withdrawal of MCF-7 cells grown as estrogen-dependent tumors in nude mice led to decreased TGF α mRNA. Recent studies with antibodies directed against either TGF α or its receptor have reported growth suppression of MCF-7 cells grown as anchorage independent colonies or as estrogen stimulated, high density monolayer cultures (7). TGF α is currently one of the most likely growth factors to exert a positive autocrine effect in breast cancer.

Is TGF α a tumor-specific growth factor? It is now known that normal human mammary epithelial cells rapidly proliferating in culture also secrete high quantities of TGF α and produce its expected mRNA (71). Thus, TGF α may contribute to growth control processes of normal as well as malignant breast tissue. Such a possibility would imply that the response of breast tissue to TGF α or EGF rather than the absolute level of its production might distinguish the cancer from normal.

D. Other Growth Factors

Additional stimulatory activities have been isolated from mammary tumor systems. For example, using a SW13 human adrenal carcinoma soft agar cloning assay, Swain isolated a factor with an apparent molecular weight of 60 kd which is secreted in large quantities by the hormone-independent breast cancer cell line MDA-MB-231 (70). It has not been determined if this factor is distinct from or related to mammary derived growth factor (4), endothelial cell growth factor (61) and fibroblast growth factor. It is not hormonally regulated in hormone responsive cell lines.

In addition to autocrine stimulated growth, estrogens may stimulate the production of growth factors in other estrogen responsive tissues which could then directly stimulate the proliferation of mammary tissue including hormone responsive breast cancer cells. A series of peptides, estromedins, have been purified from uteri and pituitaries which are mitogenic in vitro for MCF-7 cells (14,34). Estrogen-potentiating factor has been characterized as a pituitary growth factor for estrogen responsive breast cancer cells and is secreted by normal pituitary cells, as well as, pituitary tumors. Unrelated to known pituitary hormones, this factor potentiates growth of estrogen responsive breast cancer cells in the presence of estrogen in vivo and in vitro (14). No direct evidence suggests that these peptides function in vivo as estromedins.

E. Transforming Growth Factor Beta

Transforming growth factor β (TGF β) is a multifunctional 25 kd peptide that controls proliferation, differentiation, and other functions in many cell types. Many cells synthesize TGF β and essentially all of them have specific receptors for this peptide. TGF β regulates the action of many other peptide growth factors and determines the positive or negative direction of their action (66).

Human breast cancer cells secrete TGF β. However, antiestrogens and glucocorticoids, both inhibitory to MCF-7 cells, may inhibit cell growth by increasing TGF β levels and by lowering TGF α and IGF-I levels. Furthermore, several MCF-7 clones, as well as estrogen-independent breast cancer cell lines are inhibited by the exogenous addition of TGF β (43). This growth inhibition is reversed in the presence of a polyclonal antibody directed against TGF β (43). Interestingly, TGF β secretion is inhibited by treatment of MCF-7 cells with growth stimulatory estradiol and insulin. The mechanism of TGF β induction is not yet fully defined; it is not at the steady state mRNA level. Control may involve both protein synthesis and conversion of a latent form to an active form of TGF β (43). Breast cancers sometimes exist as mixtures of receptor positive and negative tumor cells. Since breast cancers do not necessarily become TGF β unresponsive as they become antiestrogen unresponsive, TGF β may act in such mixed tumors to make antiestrogens more effective than might otherwise be expected based only upon blockade of estrogen action. We have observed that in LY2 cells, an MCF-7 variant stepwise selected for antiestrogen resistance (9), TGF β is no longer induced by antiestrogen, but the cells still retain the TGF β response and TGF β receptors (43). In addition to TGF β, Gaffney and coworkers have found a 68 kd growth inhibitor for MCF-7 cells in bovine serum (21).

F. In vivo Effects of Growth Factors

The MCF-7 human breast cancer cell line requires the presence of estrogen to form tumors in the nude mouse model. As reviewed above, these cells, in vitro, produce a variety of peptide growth factors in response to treatment with estrogen. The next question then is whether these growth factors, produced and secreted into the media after estrogen stimulation in vitro, can replace the requirement for estrogen and support tumor formation in vivo. When the biological activity of serum free conditioned media from estrogen treated MCF-7 cells was tested using Alzet mini-fusion pumps in nude mice, it contained sufficient growth activity to stimulate tumor growth in nude mice, thus partially replacing estradiol (16). However, while early tumor growth was indistinguishable from estrogen induced growth, conditioned media induced tumors occurred at a lower incidence, were smaller, and tended to regress after two weeks despite reimplantation of infusion pumps every 7 days. When the conditioned media was replaced with purified EGF, tumor incidence rose above controls in the absence of estradiol, again suggesting that growth factors may act as estrogen-induced "second messengers" in estrogen induced growth in human breast cancer.

G. RAS Oncogenes and Hormonal Autonomy

Many patients initially present with hormonally responsive breast cancers. Following endocrine therapy, the breast cancer generally becomes hormone unresponsive. In an attempt to develop a model system to study the conversion of a hormone-responsive to a hormone-independent cell line, the tumor-inducing portion of the Harvey sarcoma virus, v-H-ras, was transfected into MCF-7 cells (39). The MCF-7$_{ras}$ cells contain stable integrated v-H-ras in their DNA, had 5-6 fold higher levels of mRNA as in control cells and had detectable phosphorylated v-H-ras pp21 protein. Growth of MCF-7$_{ras}$ cells was blunted in response to estrogen and antiestrogen treatment in vitro, but the transfected cells were tumorigenic in the absence of estrogen in 86% of inoculated female oophorectomized nude mice (39). MCF-7$_{ras}$ cells also expressed increased levels of the laminin receptor on their surfaces (2) and produce several unique intracellular proteins (77). In addition, MCF-7$_{ras}$ cells also exhibited increased levels of intracellular signalling as increased rates of phosphatidyl inositol turnover, analogous to estradiol treatment of MCF-7 cells (20). Increased phosphatidyl inositol turnover results in the release of the second messengers diacylglycerol and inositol triphosphate which in turn stimulate protein kinase C activity and Ca++ flux in the cells.

Growth factor secretion was also studied in MCF-7$_{ras}$ cells (18). Conditioned media prepared from MCF-7$_{ras}$ cultures as compared to control transfectant media contained 3-4 fold elevated levels of radioreceptor assayable and bioactive TGF α. A single peak of TGF α-like activity was eluted at an apparent molecular weight of 30 kd from MCF-7$_{ras}$ conditioned media. Also, secretion of immunoreactive IGF-I and TGF β were augmented 3-4 fold in MCF-7$_{ras}$ cells but PDGF secretion was not elevated.

In general, activation of ras brings about phenotypic and tumorigenic changes in human breast cancer cells, some of which are also induced by estrogens. The cells retaine the capacity to bind estrogen and respond to estrogens as shown by estrogenic induction of the progesterone receptor. Thus, ras gene transfection bypasses estrogen activation of the transformed phenotype but apparently induces a phenotype which appears to be similar but not identical to the estradiol induction pathway. Future studies will more clearly define the similarities and differences between estrogenic and v-H-ras induced malignant progression of MCF-7 cells.

Summary and Future Prospects

In summary, we have described many of the cellular events induced by estrogen in human breast cancer cells. In particular, we have emphasized evidence pointing to estrogen, polypeptide growth factors and their receptive receptors as important regulators of breast cancer cell lines in vitro as model systems. It is evident that a number of peptide growth factors play a mediating or modulating role in breast cancer cell proliferation. At present, data from a number of laboratories suggests that TGF α might be an important autostimulatory and TGF β an autoinhibitory component of growth regulation. Roles of IGF-I, PDGF and a variety of other secreted growth factors are less clear. We have also presented evidence implicating protooncogene expression in the malignant progression of normal tissue to cancer. In particular, using ras gene transfection experiments, it has been possible to induce MCF-7 and other breast epithelial cells to secrete increased amounts of growth factors and to become partially autonomous from estrogen controls.

Current efforts to control hormone responsive tumors, centered around inhibition of steroidogenesis or interference with hormone/receptor binding, universally fails when hormone independent clones emerge. As evident from in vitor studies, one mechanism by which tumors may become hormone unresponsive is by constituitively activating autocrine growth pathways normally under hormonal control. The elucidation of autocrine growth pathways, as described above, has created a wide array of new therapeutic approaches aimed at disrupting the hormonal cellular response.

With more complete knowledge of the mechanisms of regulation controlling breast cancer cell growth and malignant progression, modes of experimental breast cancer therapy could be designed to interrupt growth stimulation by TGF α, enhance the growth inhibition by TGF β or to block the paracrine effects of secreted growth factors on surrounding endothelial cells and stroma. A major obstacle in utilizing such antigrowth factor strategies in experimental therapy is a lack of knowledge of the role of growth factors in the proliferation and function of normal tissues. Such approaches have succeeded in experimental tumor systems utilizing antigrowth factor and antireceptor antibodies (12,51,52,54). Future chemotherapy approaches might utilize anti-receptor conjugates with toxins either in direct patient treatment or in removal of tumor cells from bone marrow specimens in vitro prior to reimplantation into heavily irradiated patients. These types of approaches have also begun to succeed in experimental animals (57,52).

REFERENCES

1. Aitken SC, Lippman ME (1982). Hormonal regulation of net DNA synthesis in MCF-7 human breast cancer cells in tissue culture. Cancer Res 42:1727-1735.

2. Albini A, Graf JO, Kitten T, Kleinman HK, Martin GR, Veillette A, Lippman ME (1986). Estrogen and v-ras ^{H} transfection regulate the interactions of MCF-7 breast carcinoma cells to basement membrane. Proc Natl Acad Sci (USA) 83:8182-8186.

3. Allegra JC, Lippman ME (1978). Growth of a human breast cancer cell line in serum free and response to cytotoxic chemotherapy in patients with metastatic breast cancer. Cancer Res 83:3823-3829.

4. Bano M, Salomon DS, Kidwell WR (1985). Purification of a mammary derived growth factor from human milk and human mammary tumors. J Biol Chem 260:5745-5752.

5. Barrack ER, Coffey DS (1980). The specific binding of estrogen and androgens to the nuclear matrix of sex hormone responsive tissues. J Biol Chem 255:7265-7275.

6. Bates SE, McManaway ME, Lippman ME, Dickson RB (1986). Characterization of estrogen responsive transforming activity in human breast cancer cell lines. Cancer Res 46:1707-1713.

7. Bates SE, Davidson NE, Valverius E, Dickson RB, Kudlow JE, Tam JP, Freter CE, Lippman ME, Salomon D (1987). Expression of transforming growth factor alpha mRNA in human breast cancer: regulation by estrogen. Cell, submitted

8. Bringman TS, Lindquist PB Derynck R (1987). Different transforming growth factor a species are derived from a glycosylated and palmit-oylated transmembrane precursor. Cell 48:429-440.

9. Bronzert DA, Greene GL, Lippman ME (1985). Selection and characterization of breast cancer cell line resistant to the antiestrogen LY 117018. Endocrinology 117:1409-1417.

10. Bronzert DA, Pantazis P, Antonades HN, Kasid A, Davidson N, Dickson RB, Lippman ME (1987). Synthesis and secretion of PDGF-like growth factor by human breast cancer cell lines. Roc Natl Acad Sci (USA), in press.

11. Conneely OM, Sullivan WP, Tuft DO, Birnbaumer M, Cook RG, Maxwell BL, Zarucki-Schulz T, Greene GL, Shrader WT, O'Malley BW (1986). Molecular cloning of the chicken progesterone receptor. Science 233:767-770.

12. Cuttita F, Carney DN, Mulshine J, Moody TW, Fedorko J, Fisher A, Minna J (1985). Bombesin-like peptides can function as autocrine growth factors in small lung cancer. Nature 316:823-827.

13. Davidson NE. Bronzert DA, Chambon P, Gelmann EP, Lippman ME (1986). Use of two MCF-7 cell variants to evaluate the growth regulatory potential of estrogen-induced products. Cancer Res 46:1904-1908.

14. Dembinski TC, Leung CKH, Shiu RPC (1985). Evidence of a novel pituitary factor that potentiates the mitogenic effect of estrogen in human breast cancer cells. Cancer Res 45:3083-3089.

15. Derynck R, Roberts AB, Winkler MC, Chen EY Goeddel DV (1984). Human transforming growth factor-a: precursor, structure, and expression in E. coli. Cell 38:287-297.

16. Dickson RB, McManaway M, Lippman ME (1986). Estrogen induced growth factors of breast cancer cells partially replace estrogen to promote tumor growth. Science 232:1540-1543.

17. Dickson RB, Huff KK, Spencer EM, Lippman ME (1986). Induction of epidermal growth factor-related polypeptides by 17β estradiol in MCF-7 human breast cancer cells. Endocrinology 118:138-142.

18. Dickson RB, Kasid A, Huff KK, Bates S, Knabbe C, Bronzert D, Gelmann EP, Lippman ME (1987). Activation of growth factor secretion in tumorigenic states of breast cancer induced by 17β-estradiol or v-ras^H oncogene. Proc Natl Acad Sci (USA) 84:837-841.

19. Dickson RB, Lippman ME, (1987). Estrogenic regulation of growth and polypeptide growth factor secretion in human breast carcinoma. Endocr Rev 8:29-43.

20. Freter CE, Lippman ME, Gelmann EP (1986). Hormonal effects on phosphatidyl inositol (PI) turnover in MCF-7 human breast cells. Proceedings fo the American Association for Cancer Research, Los Angeles, CA.

21. Gaffney EV, Grimwald MA, Pigott DA, Dell'Aguila M (1980). Inhibition

of growth of human breast cancer cell line MCF-7 by serum derived calcium chloride clotted plasma. J Natl Cancer Inst 65:1215-1219.

22. Gazit A, Igarashi H, Chiu I, Srinivasan A, Yaniv A, Tronick SR, Robbins KC, Aaronson SA (1984). Expression of the normal human sis/PDGF-2 coding sequence induces cellular transformation. Cell 39:89-97.

23. Giguere V, Hollenberg SM, Rosenfield MG, Evans RM (1986). Functional domains of the human glucocorticoid receptor. Cell 46:645-652.

24. Goustin AS, Leof EB, Shipley GD, Moses HL (1986). Growth factors and cancer. Cancer Res 46:1015-1029.

25. Greene GL, Gilna P, Waterfield M, Baker A, Hort Y, Shine J (1986). Sequence and expression of human estrogen receptor complementary DNA. Science 231:1150-1154.

26. Greene S, Walter P, Kumar V, Krust A, Bornert JM, Argos P, Chambon P (1986). Human oestrogen receptor cDNA: Sequence, expression and homology to v-erb-A. Nature 320:134-139.

27. Hollenberg SM, Weinberger C, Ong ES, Cerelli G, Orol A, Lebo R, Thompson ED, Rosenfield MG, Evans RM (1985). Primary structure and expression of a functional human glucocorticoid receptor cDNA. Nature 318:635-664.

28. Horwitz KB, McGuire, WL (1978). Estrogen control of progesterone receptor in human breast cancer. J Biol Chem 253:2223-2228.

29. Huff KK, Kaufman D, Gabbay KH, Spencer EM, Lippman ME, Dickson RB (1986). Human breast cancer cells secrete an insulin-like growth factor I related polypeptide. Cancer Res 46:4613-4619.

30. Huff KK, Lippman ME (1984). Hormonal control of plasminogen activator secretion in ZR-75-1 human breast cancer cells in culture. Endocrinology 114:1665-1671.

31. Huff KK, Knabbe C, Kaufman D, Lippman MEm Dickson RB (1987). Hormonal regulation of insulin-like growth factor I (IGF-I) secretion from MCF-7 human breast cancer cells. J Cellular Biochemistry Supplement 11A:28.

32. Huff KK, Knabbe C, Lindsay R, Lippman ME, Dickson RB (1987). Multihormonal regulation of insulin-like growth factor I related

protein in MCF-7 human breast cancer cells. Molecular Endocrinology, submitted.

33. Ignotz RA, Kelly B, Davis RJ, Massague J (1986). Biologically active precursor for transforming growth factor type a released by retrovirally transformed cells. Proc Natl Acad Sci (USA) 83:6307-6311.

34. Ikeda T, Lui QF, Danielpour D, Officer JB, Ho M, Leland FE, Sirbasku DA, (1982). Identification of estrogen inducible growth factors (estromedins) for rat and human mammary tumor cells in culture. In Vitro 18:961-975.

35. Jakesz R, Smith CA, Aitken S, Huff KK, Schuette W, Shackney S, Lippman ME (1984). Influence of cell proliferation and all cycle phases on expression of estrogen receptor in MCF-7 breast cancer cells. Cancer Res 44:619-625.

36. Jakowlew SB, Breathnack R, Jeltsch J, Masiakowski P, Chambon P (1984). Sequence of the pS2 mRNA induced by estrogen in the human breast cancer cell line MCF-7. Nucleic Acid Res 12:2861-2874.

37. Jeltsch JM, Krozowski Z, Quirin-Stricker C, Gronemeyer H, Simpson RJ, Garnier JM, Krust A, Jacob F, Chambon P (1986). Cloning of the chicken progesterone receptor. Proc Natl Acad Sci (USA) 83:5424.

38. Jansen M, Van Schaik FMA, Ricker AT, Bullock B, Woods PE, Gabbay KH, Nussbaum AL, Sussenback JS, Vander Branch JR (1983). Sequence of cDNA encoding human insulin-like growth factor I precursor. Nature 306:609-611.

39. Kasid A, Lippman ME, Papageorge AG, Lowy DR, Gelmann EP (1985). Transfection of v-ras^H DNA into MCF-7 cells bypasses their dependence on estrogen for tumorigenicity. Science 228:725-728.

40. Kasid A, Davidson N, Gelmann E, Lippman ME, (1986). Transcriptional control of thymidine kinase gene expression by estrogen and antiestrogen in MCF-7 human breast cancer cells. J Biol Chem 261:5562-5567.

41. Kaufman U, Zapf J, Torrett B, Froesch ER (1977). Demonstration of a of a specific serum carrier protein of nonsuppressible insulin-like activity in vitro. J Clin Endocrinol Metab 44:160-166.

42. King WJ, Greene GL (1984). Monoclonal antibodies localize estrogen receptor in nuclei of target cells. Nature 307:745-749.

43. Knabbe C, Wakefield L, Flanders K, Kasid A, Derynck R, Lippman M, Dickson RB (1987). Evidence that TGFβ is a hormonally regulated negative growth factor in human breast cancer. Cell 48:417-429.

44. Krust A, Green S, Argos P, Kumar V, Walter P, Bornert JM, Chambon P (1986). The chicken oestrogen receptor sequence: Homology with v-erb A and the human oestrogen and glucocorticoid receptors. IMBO J 5:891-897.

45. Kumar V, Green S, Staub A, Chambon P (1986). Localization of the oestradiol-binding and putative DNA-binding domains of the human oestrogen receptor. EMBO J 5:2231-2236.

46. Levine RM, Rubalcaba E, Lippman ME, Cowan KH (1985). Effects of estrogen and tamoxifen on the regulation of dihydrofolate gene expression in a human breast cancer cell line. Cancer Res 45:1-7.

47. Liotta L (1985). Tumor invasion and metastases: Role of the extracellular matrix. Proc Am Assoc Cancer Res 26:385-386.

48. Lippman ME, Bolan G, Huff KK (1972). Interactions of antiestrogens with human breast cancer in long term tissue culture. Cancer Treat Rep 60:1421-1430.

49. Lippman, ME, Bolan G, Huff KK (1976). The effects of estrogens and antiestrogens on hormone responsive human breast cancer in long-term tissue culture. Cancer Res 36:4595-4601.

50. Lippman ME (1985). Endocrine responsive cancers of man. In Williams RH (ed):Textbook of Endocrinology, WB Sanders Co, Philadelphia 1309-1326.

51. Masui H, Kawamoto T, Sato JD, Wolf B, Sato G, Mendelsohn J (1984). Growth inhibition of human tumor cells in athymic mice by anti-epidermal growth factor receptor monoclonal antibodies. Cancer Res 44:1002-1007.

52. Masui H, Morogama T, Mendelsohn J (1986). Mechanism of antitumor activity in mice for anti-epidermal growth factor receptor monoclonal antibodies with different isotypes. Cancer Res 46:5592-5598.

53. May FEB, Westley BR (1986). Cloning of estrogen-regulated messenger RNA sequences from human breast cancer cells. Cancer Res 46:6034.

54. Mendelsohn J, Masui H, MacLeod C (1985). Anti-EGF receptor monoclonal antibody 528 inhibits proliferation of a subset of human tumor cells in xenografts. Proc Am Assoc Cancer Res 26:287.

55. Neufeld EF, Ashwell G (1980). Lennarz WJ (ed): The Biochemistry of Glycoproteins and Proteoglyeans. Plenum Press, New York, pp.241-266.

56. Pardoll DM, Vogelstein B, Coffey DS (1980). A fixed site of DNA replication in eucaryotic cells. Cell 19:527-536.

57. Pastan I, Willingham MC, Fitzgerald DP (1986). Immunotoxins. Cell 47:641-648.

58. Dedman J, Tam J (1986). Immunological detection and quantitation of α transforming growth factors in human breast carcinoma cells. Breast Cancer Research and Treatment 7:201-210.

59. Puca GA, Sica V, Nola E (1974). Identification of a high affinity nuclear acceptor site for estrogen receptor of calf virus. Proc Natl Acad Sci (USA) 171:979-983.

60. Robinson SI, Nelkin BD, Vogelstein B (1985). The ovalbumin gene is associated with the nuclear matrix of chicken oviduct cells. Cell 28:99-106.

61. Rosenthal A, Lindquist PB, Bringman TS, Goeddel DV, Derynck R (1986). Expression in rat fibroblasts of a human transforming growth factor -a cDNA results in transformation. Cell 46:301-309.

62. Salomon DS, Zwiebel JA, Bano M, Losonczy, Felnel P, Kidwell WR (1984). Presence of transforming growth factors in human breast cancer cells. Cancer Res 44:4069-4077.

63. Schreiber AB, Kenney J, Kowalski J, Thomas KA, Gimenez-Gallego G, Rios-Candelore M, DiSalvo J, Bamitault D, Courty J, Courtois Y, Moemer M, Loret C, Burgess WH, Mehlman T, Friesel R, Johnson W, Maciag T (1985). A unique family of endothelial cell polypeptide mitogens: The antigenic and receptor cross-reactivity of bovine endothelial growth factor and eye-derived growth factor II. J Cell Biol 101:1623-1626.

64. Schuh S, Yamemoto W, Brugge J, Bauer VJ, Riehl RM, Sullivan WP, Toft DO (1985). A 90,000 dalton binding protein common to both steroid receptors and the rous sarcoma virus transforming protein $pp60^{V-SRC}$. J Biol Chem 260:14292:14296.

65. Spelsberg TC, Webster RA, Pikler GM (1976). Chromosomal proteins regulate steroid binding to chromatin. Nature 262:65-67.

66. Sporn MB, Roberts AB, Wakefield LM, Assoian RK (1986). Transforming growth factor B: Biologic function and chemical structure. Science 233:532-534.

67. Stoppell MP, Tacchetti C, Cabellis MV, Corti A, Hearing VJ, Cassani G, Appella E, Blasi F (1986). Autocrine saturation of prourokinase receptor on human A431 cells. Cell 45:675-684.

68. Stoscheck CM, King Jr LE (1986). Role of epidermal growth factor in carcinogenesis. Cancer Res 46:1030-1037.

69. Strobl JS, Thompson EB (1985). Mechanism of steroid hormone action. In Auricchio, F. (ed): Sex Steroid Receptors, Field Educational Helia Acta Medica, Rome, pp9-36.

70. Swain S, Dickson R, Lippman ME (1986). Anchorage independent epithelial colony stimulating activity in human breast cancer cell line. Proc Amer Assoc Cancer Res 27:844.

71. Valvarius E, Bates SE, Salomon DS, Stampfer M, Clark R, McCormick F, Lippman ME, Dickson RB (1987). Production of transforming growth factor α by human mammary epithelial cells and sublines. Proceedings of the Annual Meeting of the Endocrine Society, Indianapolis, Ind in press.

72. Vitella ES, Uhv JW (1985). Immunotoxins: redirecting nature's poisons. Cell 41:653-665.

73. Walter P, Green S, Greene G, Krust A, Bornert JM, Jeltsch JM, Straub A, Jensen E, Scrace G, Waterfield M, Chambon P (1985). Cloning of the human estrogen cDNA. Proc Natl Acad Sci (USA) 82:889-893.

74. Weinberger C, Thompson CC, Ong ES, Lebo R, Gruel DJ, Evans RM (1986). The c-erb-A gene encodes a thyroid hormone receptor. Nature 234:641-646.

75. Welshons WV, Leiberman ME, Gorski J (1984). Nuclear localization of unoccupied estrogen receptors. Nature 307:747-749.

76. Westley B, Rochefort H (1980). A secreted glycoprotein induced by estrogen in human breast cancer cell lines. Cell 20:353-362.

77. Worland PJ, Bronzert DA, Dickson RB, Lippman ME, Thorgeirsson SS, Wirth PJ (1987). Celular polypeptide patterns of MCF-7 and transfected MCF-7 (ras) human brest cancer cells. J Cellular Biochemistry Supplement 11A, p.84 (Abstract #A293).

PEPTIDE HORMONES AND RECEPTORS

CELLULAR ENDOCRINOLOGY AND CANCER

Gordon H. Sato

W. Alton Jones Cell Science Center, Inc.
10 Old Barn Road
Lake Placid, New York 12946

It is a great honor and a great challenge to be placed on the program after Dr. Huggins. A few years ago I wrote him a letter and sent him an essay I had written entitled "Towards an Endocrinology of Cancer". I got back a very kind, encouraging letter which I valued very much.

My laboratory over the years has been involved in or has contributed to the solution of two problems in the technique of tissue culture. The first problem was the lack of differentiated function in cell culture lines. This situation persisted for over fifty years after the initial development of the tissue culture technique by Ross Harrison, and it was a major stumbling block to the realization of the dream people had when the tissue culture technique was first described. That dream was to be able to take animals apart cell type by cell type, study them in a controllable environment and find out how the animal as a whole is put together. The second problem was to determine why it was always necessary to include serum or some biological fluid in the media of cells growing in culture. It turns out that these two problems are connected. The solution of one is linked to the solution of the other. Some thirty years ago when we started this work, both of these were daunting problems. At the time everybody thought that cells in culture underwent a process of dedifferentiation whereby the parenchymal cells of the tissue origin gradually change into fibroblasts. We showed that what was actually happening was that the parenchymal cells or functional cells of the tissue of origin weren't changing into fibroblasts but that fibroblasts, although few in number in the original inoculum, were overgrowing the rest of the cells (Figure 1). This of course is obvious today. If you want to culture epithelial cells you have to worry that fibroblasts don't overgrow the epithelial cells. However, at the time we did this work, this dogma was so firmly entrenched that I got a letter from the American Cancer Society stating that I was propounding nonsense, and that I need not bother reapplying to that particular organization for a grant again. That was how strong feelings were at that time. However, once we realized that the problem in getting functionally differentiated cell lines was a matter of growing the right kind of cell, we decided to turn to microbiology and the technique of enrichment culture. To show you how these problems are connected, today we know that the reason for selective overgrowth of fibroblasts in serum containing medium is that platelet derived growth factor selectively encourages the growth of fibroblasts. So we turned to a technique we thought of as enrichment culture and to one of the great pioneers of endocrine oncology, Jacob Furth, and obtained some functionally differentiated tumors from him. We took these tumors and put them into culture. After a while, we put the cells into animals and the resulting tumors back into culture. We kept this process going until we got a tumor which we could grow readily in culture. In this way we were able to establish a number of functionally differentiated lines derived from transplantable tumors (Figure 2). These included adrenocortical cells that produced steroids in response to ACTH, ACTH producing pituitary cells, growth hormone producing pituitary cells, glial cells, neuroblastoma cells and teratocarcinoma cells which could differentiate in culture. Part of the problem of establishing functionally differentiated cultures at this time was to apply fairly rigorous biochemical characterization which was not common among scientists who grew cells in culture. Our adrenocortical cultures secreted a substance in response to ACTH which had a maximal absorption at 240 millimicrons. We had to learn steroid chemistry to prove that this material producing optical density at 240 was actually a steroid. That turned out to be a terribly difficult job because it was a steroid that had just

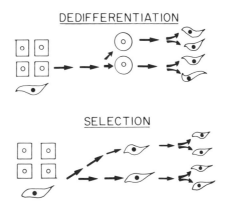

FIGURE 1 - This is a graphical representation of the two hypotheses to explain why all cell culture stains previous to the 1960's had the characteristics of fibroblasts. In this case, liver is taken as the illustrative example. The groups of cells on the left represent the cell population taken from the liver to be placed in culture. It consists mainly of the parenchyma or organ-specific cells of the liver (square cells) and to a small extent of connective tissue fibroblasts (elongated cells).

Both hypotheses explain why the final tissue culture population (group of cells on the right) are fibroblastic. According to the dedifferentiation hypothesis, parenchymal cells growing in culture are gradually converted into fibroblasts. According to the selection hypothesis, the minor component of fibroblasts in the initial inoculum outgrow the parenchymal cells so that the final culture population consists of fibroblasts derived from preexisting fibroblasts. To determine which explanation was correct, antisera against parenchyma and antisera against fibroblasts were prepared. Cells, freshly isolated from the liver, were treated with the antisera for brief periods of time, and then placed in culture. The dedifferentiation hypothesis predicts that killing the parenchyma of the inoculum with antisera would block subsequent culture growth because it is the parenchyma which grow and become fibroblasts. The selection hypothesis predicts that killing the fibroblasts, which are only a minor component of the initial population, would block subsequent growth because it is these cells that generate the final population. The experimental results were consistent with the selection hypothesis.

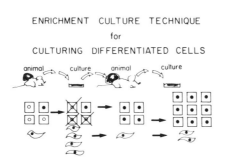

FIGURE 2 - The method of alternately passaging differentiated tumor cells through animals and culture is graphically presented. The original tumor cell population is represented by the group of cells on the left. It consists of the differentiated tumor cells (square cells) and fibroblasts. When this cell population is put into culture, the fibroblasts proliferate to a great extent while the tumor cells die. A few exceptionally hardy tumor cells (represented by cells with dark nuclei) survive the initial culture period. When the cultures are injected into animals, a tumor is generated from these hardy tumor cells which have been selected for their ability to survive the culture conditions. The fibroblasts contribute little to the growth of the tumor because they are of noncancerous cells. The culture-derived tumors grow more readily in culture than the original tumor because of the previous selection. From these cultures; it is now easier to isolate pure strains of functionally differentiated tumor cells. In this way adrenal cortical (steroid-secreting, ACTH-responsive), pituitary (ACTH-secreting), pituitary (prolactin- and growth hormone-secreting), Leydig (steroid-secreting), glial (S100 protein-containing), neuroblastoma (electrically-active, neurotransmitter-synthesizing), teratoma (capable of differentiation), and melanoma (pigmented) cell lines were established.

been discovered a year before. Since it was a derivative of progesterone with a 20 alpha hydroxyl group, it was difficult to identify.

The fact that we could culture various cells like adrenocortical cells and pituitary cells which perform functionally differentiated functions of the tissue origin led us to attack another problem. Why was it that no cells in culture exhibited a growth response to hormones? In animals, we know that the stimulation of proliferation is caused by hormones. The uterus proliferates in response to steroids and the ovary responds to gonadotrophins. Why were cells in culture not giving a growth response to hormones? In our search for answers we developed a model system based on the work of Biskind and Biskind, two pathologists who some twenty years previously had established ovarian tumors by taking an animal, removing its ovary, and implanting portions of that ovary in the spleen. This ovarian implant would produce steroid hormones. These steroid hormones would not go into the general circulation but to liver capillaries via the hepatic portal vein. The liver would inactivate these steroids. The pituitary, sensing no steroids, would hypersecrete gonadotrophins causing the implant in the spleen to grow very large. Presumably, the growth of this huge mass of ovarian cells was driven by gonadotrophins. We established such cultures and cloned them. The clones would only grow in animals if the animal was ovariectomized and the cells were implanted in the spleen. However, we could not show that there was a hormonal dependence in culture. This was a great disappointment because after two years of work to reach that stage, we finally realized the serum in the tissue culture medium was contributing hormones. After we developed various ways to strip the hormones from serum, we found the hormonal dependence was readily demonstrable. Luteinizing hormone which we expected to drive the proliferation of these cells turned out not to be the active factor. Instead, it was a contaminating hormone which today we know as FGF, but at that time we called it OGF. Eventually we stripped the serum of hormones and learned that cells responded to a great number of hormones, or growth factors, some of which had not seen before. So, Izumi Hayashi, a graduate student at the time, followed this up by seeing if she could replace the serum with combinations of hormones. We used one of the cell lines we had developed, GH_3 pituitary cells, that made growth hormone to prolactin. She was able to show that she could replace the serum in their medium with a combination of factors, most of which were classical hormones (Table 1). From this we learned that serum was providing actually a small number of factors. This is important because serum is a very complicated biological mixture. If you apply two dimensional electrophoresis to serum, you'll find approximately 1,000 different proteins and peptides. These could be replaced in the medium of GH_3 cells with 7 purified substances. The other conclusion of this experiment is that the endocrinology of any kind of cell is very complicated. The advantage of the tissue culture approach is evident when you list the origin of these factors and the kinds of ablations one would have to perform to demonstrate hormonal dependence. To show this kind of hormonal dependence in a whole animal is impossible, so the best way is through serum-free tissue culture.

Using defined medium without serum also allows one to culture cells one could not culture before. One of the first examples was the work of the laboratory of Hayden Coon. They were able to establish rat thyroid follicular cells which function in the sense that they produce thyroglobulin and respond to thyroid stimulating hormone. The way you establish these cells in culture is that you use no serum, but you use a mixture of various hormones. The point is that if one took a thyroid gland and put it into serum containing medium, one would never establish functional thyroid follicular cells, but if they are established from the outset in primary culture in defined medium, one always get functional thyroid cells. The other thing that is very noteworthy about these cells is that they don't undergo the loss of proliferative capacity; they don't undergo senescence in culture. I think the whole question of senescence in cell culture is going to have to be reexamined as people more and more often establish cultures using defined media.

Another cell that one could never establish by taking a gland and placing it into a serum-containing medium is parathyroid cells. The laboratory of Hayden

HORMONAL REQUIREMENTS
of

GH_3 { Growth Hormone
Prolactin Secreting
Pituitary Cells }

	ORGAN SOURCE
1. INSULIN	Pancreas
2. TRANSFERRIN	Liver
3. THYROXIN	Thyroid
4. PARATHYROID HORMONE	Parathyroid
5. THYROID STIMULATING HORMONE RELEASING HORMONE	Hypothalamus (Brain)
6. FIBROBLAST GROWTH FACTOR	Pituitary
7. SOMATOMEDIN C	Liver

TABLE 1 — *The hormonal and factor requirement of GH_3 pituitary cells for growth in serum-free medium are presented here with the organ source of each component in parentheses. In view of the immense complexity of serum, it is surprising how so few a number of defined substances can replace serum. It is also surprising that the cells give a positive growth response to so many hormones. The serum-free technology is also a powerful tool for revealing hormonal dependencies. The organs listed in parentheses are the organs that would have to be ablated to show these responses by classical endocrine experiments.*

Coon has established parathyroid cells which produce parathyroid hormone and give the appropriate response to calcium concentrations in the medium. And again these cells do not senesce. So that is a very curious thing. If you establish primary cultures in defined medium, the cells don't undergo what is called the Hayflick phenomenon.

There have been talks about hormones and cancer, and I would just like to give my point of view on this. In agreement with Dr. Hankins, I would say that virtually all cells require hormones for survival and growth including cancer cells. Their dependence is complicated, and because it is complicated it seems to me that they are vulnerable. The approach we have been taking is to use monoclonal antibodies to hormone receptors, namely the EGF receptor. In animals we can show that we can block the tumor growth of A431 cells in nude mice if we administer monoclonal antibodies to the EGF receptor. I think one of the approaches that we will be taking in the future is to block all the hormone receptors of a cancer cell which can be identified. The reason why single hormone deletions are not effective is because these cells respond to hormones in a peculiar way. If you delete any single one, you don't stop growth, you slow it down. But if you delete all of them, you will kill the cells. I think that if one can develop monoclonal antibodies to all the receptors, one can achieve a kind of specificity because the combination of hormones that any kind of cancer cell requires is different from that of other cells in the body. One must keep in mind that many of these cancer cells are producing autocrine factors, and they are very difficult to detect because they are produced in very low concentrations. The cell biology experiments must be carefully designed in order to discover these autocrine factors. If one can block all the receptors for autocrine, paracrine and endocrine hormones then one may be able to make some progress.

The point of cell culture of functionally differentiated cells in hormonally defined media is that it is a new approach to organ physiology. How complexes of hormones interact in bringing about finely controlled physiological responses can best be studied in this way.

REFERENCES

Barnes D, Sato G (1979). Serum-free growth of cells in culture. Analytical Biochemistry, submitted for publication.

Sato G, Augusti-Tocco G, Posner M (1970). Hormone-secreting and hormone-responsive cell cultures. Recent Progress in Hormone Research 26:539.

EPIDERMAL GROWTH FACTOR

Barbara Mroczkowski and Graham Carpenter

Department of Biochemistry
Vanderbilt University School of Medicine
Nashville, Tennessee 37232 USA

INTRODUCTION

The discovery and isolation of substances involved in the control of growth and development in animal cells have begun to provide a detailed understanding of how certain mitogens mediate their biological activities. We now know that the initial interaction of growth factors with target cells involves their specific binding to cell surface receptors and the subsequent internalization of ligand-receptor complexes. It has also become evident during the past decade that many of the growth factor receptors represent bifunctional transmembrane molecules with the different functional domains exhibiting the ability to communicate with one another. The mechanism(s) by which binding of the peptide growth factor to the extracellular domain of the transmembrane receptor activates the cytoplasmic catalytic domain, and more importantly, how this extracellular signal is transduced to the nucleus to activate DNA replication remain to be elucidated.

Epidermal growth factor (EGF) is perhaps one of the most comprehensively studied peptide growth factors. The initial work on EGF was pioneered by Stanley Cohen and has laid out a foundation for our understanding at the biochemical level of how extracellular peptides produce biological responses in target cells. A historical perspective highlighting some of the major advancements in this area of research will be the focus of this review.

HISTORICAL PERSPECTIVES

In 1960, while studying a factor detected in male mouse submaxillary glands which was responsible for promoting the growth of embryonic neurons, Stanley Cohen made his initial observation that crude submaxillary gland preparations upon injection into newborn mice promoted the appearance of specific developmental processes. He observed that daily subcutaneous injections of EGF accelerated both eyelid opening and incisor eruption. The eyes of mice injected with this crude preparation began to open at six to seven days after birth; compared to 12 to 14 days for the uninjected mice. In an analogous fashion, their incisors erupted at 5-6 days instead of the normal 8-10 days (1).

Employing precocious eyelid opening as a functional assay, Cohen proceeded to isolate the factor responsible for these biological effects from murine submaxillary glands in the early 1960's (2). The factor was named epidermal growth factor (EGF) since it became evident that it was responsible for the enhancement of epidermal growth and keratinization both in vivo and in vitro using organ cultures of chick embryonic skin (3).

The development of a rapid isolation procedure for EGF from murine submaxillary glands expedited our understanding of the chemical and physical properties of EGF (4). The primary sequence of murine EGF indicates that it is a polypeptide of 53 amino acid residues with three intrachain disulfide bonds (5). This molecule is not peculiar to the mouse, but is found in all mammals including humans. Human epidermal growth factor (hEGF) has been isolated from urine (6,7). Urogastrone, a human gastric antisecretory hormone, is structurally and functionally identical to hEGF (8).

EPIDERMAL GROWTH FACTOR

Numerous proteins share sequence homology to EGF. Among these are human pancreatic secretory trypsin inhibitor (9), α-transforming growth factor (α-TGF) (10, 11), the 19 kDa early protein of vaccinia virus (12, 13, 14), and two homeotic gene products, the Drosophila Notch gene product (15) and the lin-12 gene product of the nematode Caenorhabditis (16). α-TGF and the vaccinia growth factor have been shown to be functionally equivalent to EGF in various assays. Human α-TGF, an EGF analogue produced by both fetal and transformed cells has limited homology (33-44%) with the amino acid sequences of human and mouse EGFs (10, 11, 17) and yet exhibits high affinity binding to EGF receptors and elicits the same biological responses as EGF (10, 18).

A close comparison of the sequences of EGF, α-TGF and the 19 kDa vaccinia virus growth factor reveals very little overall sequence conservation. Of the 53 residues in EGF, only 12 residues are strictly conserved and of those, six are represented by cysteine residues, indicating that the placement of disulfide bonds might govern the ability of these molecules to interact with the EGF receptor.

Analysis of cDNA clones derived from mRNA transcripts encoding mEGF has revealed that this peptide is synthesized as a large precursor molecule of 1217 residues with a hydrophobic transmembrane domain (19, 20, 21). Another interesting observation about this precursor is that it contains eight regions of partial sequence homology to mature EGF (22). Whether these EGF-related sequences are ever processed to serve a biological function is not known. A portion of the EGF precursor also exhibits limited homology with the sequence of the low density lipoprotein (LDL) receptor, suggesting perhaps that the EGF precursor and the LDL receptor both descended from a common ancestral transmembrane protein (23,24).

Radioimmunoassay analysis in conjunction with immunohistochemical data have revealed the highest levels of mEGF to be present in the male submaxillary gland with considerably lower amounts detected in kidney, stomach, paratid and pancreas (25). The human tissues reported to contain hEGF are kidney, small intestine, thyroid gland, pancreas and male submandibular gland (26). Nearly all human body secretions (milk, sweat, tears, saliva, urine) and fluids (plasma, amniotic fluid) contain EGF, though plasma levels are quite low and EGF may not be a classical circulating hormone. cDNA hybridization studies have detected the presence of EGF mRNA transcripts in the submaxillary gland, kidney, mammary gland, pancreas, duodenum, pituitary, lung and spleen (27). The function of EGF in the kidney remains to be elucidated, but is of particular interest in light of the observation that EGF in this organ is isolated as the intact precursor molecule which does not appear to be processed to the mature peptide.

THE RECEPTOR FOR EGF

In the early 1970's, two reports (28,29) demonstrated that cultured fibroblasts responded to EGF by the induction of DNA synthesis. The enhancement of DNA synthesis mediated by EGF was later corroborated in Cohen's laboratory using human fibroblasts (30, 31). These findings predicted the presence of cell surface binding sites for EGF in cultured cell lines. Binding studies utilizing ^{125}I-EGF and human fibroblasts indicated that the cell surface receptor for EGF facilitated the internalization of the mitogen into the cells where it was subsequently proteolyzed (32).

Immunofluorescence and electron microscopic studies further confirmed the biochemical data obtained with ^{125}I-EGF. Using biologically active EGF conjugated to ferritin, it was possible to visualize the clustering and internalization within endocytic vesicles of ferritin-EGF molecules bound to the receptor as well as their subsequent transport to multivesicular structures or lysozymes (33-37). Taken together, these data indicated that "down-regulation" of surface receptors for EGF was a consequence of hormone-receptor internalization. Further biochemical studies involving the metabolic labeling of cell lines containing EGF receptors and immunoprecipitation of the labelled receptor clearly demonstrated that ligand mediated internalization of EGF-receptor complexes results not only in the degradation of EGF, but also in the enhanced degradation of the receptor (38,39).

The presence of EGF receptors in various cell types (fibroblasts, glial cells, granulosa cells, endothelial and epithelial cells) has been documented (40). Receptors for EGF have also been found in a wide spectrum of mammalian tissue such as liver, placenta, skin, cornea, kidney, and brain (41-42). EGF receptors have also been detected in numerous human neoplasms including choriocarcinomas, non-neuronal brain tumors, and squamous cell carcinomas (43-45). The only group of cells that has not been reported to have receptors for EGF are cells of the circulatory system. A gene coding for an EGF receptor-related protein in <u>Drosophila</u> <u>melanogaster</u> has also been cited (46).

Numerous cell lines from diverse sources exhibit the capability of specifically binding ^{125}I-EGF. One particular cell line, the A-431 human epidermoid carcinoma cell line, has been instrumental in our understanding of receptor structure and function. This cell line is unique in the sense that it over-expresses the EGF receptor ($2-3 \times 10^6$ receptors/cell). In terms of protein purification, this cell line provided a 20-fold enrichment in starting material for the isolation of the EGF receptor molecule. Solubilization of A-431 membrane preparations and subsequent affinity purification using wheat germ lectin sepharose and EGF conjugated to an agarose gel revealed the EGF receptor to be a glycoprotein of $M_r = 170,000$ (47,48).

Further examination of the structural and functional properties of the EGF receptor disclosed the intrinsic association of a protein kinase activity within the cytoplasmic domain of the molecule. It was initially observed that membrane preparations incubated with EGF exhibited a rapid increase in protein kinase activity (49). Incubation of A-431 membrane preparations with EGF resulted in a substantial enhancement of phosphorylation of endogenous proteins, as well as exogenously added protein substrates. The major phosphorylated membrane protein observed after EGF-induced activation of the kinase exhibited a molecular weight similar to that of the EGF receptor. These early observations led to the premise that phosphorylation of membrane-associated proteins might represent one of the initial events in signal transduction regulating cell growth.

It was subsequently shown that the EGF-sensitive protein kinase activity was specific for tyrosyl residues (50) and that affinity purified receptor preparations not only retained this EGF-enhanced kinase activity, but also served as a substrate for the associated kinase (51). A number of experimental approaches were utilized to ascertain whether the EGF binding site, the kinase and substrate domains all resided within the same molecule. All three properties remained associated with the receptor after electrophoretic separation under non-denaturing conditions, as well as after immunoprecipitation with sera specific for the EGF receptor (52). More conclusive evidence was obtained using the ATP analogue 5'-p-fluorosulfonylbenzoyl adenosine (FSBA) to affinity label the ATP binding site of the kinase. These experiments demonstrated that FSBA specifically bound to purified EGF receptor preparations and irreversibly inhibited the EGF-stimulated kinase activity of the receptor (53,54). These data in conjunction with the deduced amino acid sequence derived from cDNA sequence analysis (55), support a functional model for the EGF receptor. In this model the receptor is viewed as an allosteric transmembrane enzyme with the ligand binding site residing in the extracellular domain and the enzymatic activity residing in the cytoplasmic domain. Most available data suggest the presence of one transmembrane segment in the receptor sequence. Since these initial observations were made, it has become evident that a number of other growth factor receptors also represent ligand-activated tyrosine kinases (56).

The sequence of the cDNA encoding the human EGF receptor indicates that the EGF receptor is closely related to the src gene family of protein tyrosine kinases (55). More interestingly, residues 551-1154 of the human EGF receptor exhibit greater than 90% homology with the predicted sequence of the v-erb B gene product of the avian erythroblastosis virus (AEV) (57). On a structural basis, the v-erb B protein appears to represent a truncated form of the EGF receptor which lacks most of the extracellular ligand binding domain and a small portion of the cytoplasmic carboxyl terminus. It has been proposed that the transforming properties associated with the v-erb B protein are a manifestation of constitutive activation of the tyrosine kinase activity.

Physiological substrates for the EGF receptor, or any other tyrosine kinase, remain to be identified. Autophosphorylation of the EGF receptor represents the only example of a substrate with a known function to be stoichiometrically phosphorylated in an EGF dependent manner both in vivo and in vitro. However, the functional significance of autophosphorylation of the EGF receptor remains unclear and a similar lack of understanding exists with respect to exogenous substrates for this kinase activity. While several proteins are known to be phosphorylated both in vitro and in vivo on tyrosyl residues in an EGF-dependent manner, the functional significance of their covalent modifications remains unknown.

In addition to the involvement of tyrosyl kinase activity in normal and abnormal cellular growth, the influence of other signal transducing systems on cellular growth also is being examined. It is known that, in certain cells, phosphatidylinositol turnover increases and that significant amounts of IP3 are produced in response to EGF binding to its receptor (58). Furthermore, activation of protein kinase C by hydrolysis products of phosphatidylinositol may be responsible for the phosphorylation of the EGF receptor by protein kinase C at threonine 654. Phosphorylation of this threonine residue has been implicated in modulating EGF receptor function (59,60).

In summation, we currently have a clear understanding of the primary events involved when a specific ligand binds to its respective receptor. However, the nature and pathway of the signals involved in regulating nuclear transcription and ultimately DNA replication remain elusive.

ACKNOWLEDGEMENTS

The authors thank Ms. T. Hamilton for preparation of the manuscript. Grant support is acknowledged from the National Cancer Institute, grants CA43720 and CA24071.

REFERENCES

1. Cohen, S. (1960) Proc. Natl. Acad. Sci. USA 46:302-311.
2. Cohen, S. (1962) J. Biol. Chem. 237: 1555-1562.
3. Cohen, S., and Elliott, G.A. (1963) J. Invest. Dermatol. 40:1-5.
4. Savage, C.R., Jr., and Cohen, S. (1972) J. Biol. Chem. 247:7609-7611.
5. Savage, C.R., Jr., Hash, J.H., and Cohen, S. (1973) J. Biol. Chem. 248:7669-7672.
6. Starkey, R.H., Cohen, S., and Orth, D.N. (1975) Science 189:800-802.
7. Cohen, S., and Carpenter, G. (1975) Proc. Natl. Acad. Sci. U.S.A. 72:1317-1321.
8. Gregory, H. (1975) Nature 257:325-327.
9. Hunt, L.T., Barker, W.C., and Dayhoff, M.O. (1974) Biochem. Biophys. Res. Commun. 60:1020-1028.
10. Massague, J. (1983) J. Biol. Chem. 258:13606-13620.
11. Marquardt, H., Hunkapiller, M.W., Hood, L.E., Twardzik, D.R., DeLarco, J.E., Stephenson, J.R., and Todaro, G.J. (1983) Proc. Natl. Acad. Sci. USA 80:4684-4688.
12. Blomquist, M.C., Hunt, L.T., and Barker, W.C. (1984) Proc. Natl. Acad. Sci. USA 81:7363-7367.

13. Brown, J.P., Twardzik, D.R., Marquardt, H., and Todaro, G.J. (1985) Nature 313:491-492.
14. Reisner, A.H. (1985) Nature 33:801-803.
15. Wharton, K.A., Johansen, K.M., Xu, T., and Artavanis-Tsakonas, S. (1985) Cell 43:567-581.
16. Greenwald, I. (1985) Cell 43:583-590.
17. Marquardt, H., Hunkapiller, M.J., Hood, L.E., and Todaro, G.J. (1984) Science 223:1079-1082.
18. Pike, L.J., Marquardt, H., Todaro, G.J., Gallis, B., Casnellie, J., Bornstein, P., and Krebs, E.G. (1982) J. Biol. Chem. 257:14628-14631.
19. Gray, A., Dull, T.J., and Ullrich, A. (1983) Nature 303: 722-725.
20. Scott, J. Urdea, M., Quiroga, M., Sanchez-Pescador, R., Fong, N., Selby, M., Rutter, W.J., and Bell, G.I. (1983) Science 221:236-240.
21. Pfeffer, S., and Ullrich, A. (1985) Nature 313:184.
22. Doolittle, R.F., Feng, D.F., and Johnson, M.S. (1984) Nature 307:558-560.
23. Russell, D.W., Schneider, W.J., Yamamoto, T., Luskey, K.L., Brown, M.S., and Goldstein, J.L. (1984) Cell 37: 577-585.
24. Yamamoto, T., Davis, C.G., Brown, M.S., Schneider, W.J., Casey, M.L., Goldstein, J.L., and Russell, D.W. (1984) Cell 39:27-38.
25. Byyny, R.L., Orth, D.N., and Cohen, S. (1972) Endocrinology 90:1261-1266.
26. Hirata, Y., and Orth, D.N. (1979) J. Clin. Endocrinol. Metab. 48:667-679.
27. Rall, L.B., Sott, J., Bell, G.I., Crawford, R.J., Penshow, J.D., Niall, H.D., and Coghlan, J.P. (1985) Nature 313:228-231.
28. Armelin, H. (1973) Proc. Natl. Acad. Sci. USA 70:2702-2706.
29. Hollenberg, M.D., and Cuatrecasas, P. (1973) Proc. Natl. Acad. Sci. USA 70: 2964-2968.
30. Cohen, S., Carpenter, G., and Lembach, K.J. (1975) Adv. Metab. Digord. 8:265-284.
31. Carpenter, G., and Cohen, S. (1976) J. Cell. Physiol. 88:227-237.
32. Carpenter, G., Lembach, K.J., Morrison, M., and Cohen, S. (1975) J. Biol. Chem. 250:4297-4304.
33. Haigler, H., Ash, J.F., Singer, S.J., and Cohen, S. (1978) Proc. Natl. Acad. Sci. USA 75:3317-3321.
34. Gorden, P., Carpentier, J.-L., Cohen, S., and Orci, L. (1978) Proc. Natl. Acad. Sci. USA 75:5025-5029.

35. Haigler, H.T., McKanna, J.A., and Cohen, S. (1978) J. Cell Biol. 81:382-395.
36. Haigler, H., and Cohen, S. (1979) Trends Biochem. Sci. 4:132-134.
37. McKanna, J.A., Haigler, H., and Cohen, S. (1979) Proc. Natl. Acad. Sci. USA 76:5689-5693.
38. Stoscheck, C.M., and Carpenter, G. (1984) J. Cell. Biol. 98:1048-1053.
39. Decker, S.J. (1984) Mol. Cell. Biol. 4:571-575.
40. Carpenter, G. (1981) Hanb. Exp. Pharmacol. 57:89-132.
41. O'Keefe, E., Hollenberg, M.D., and Cuatrecasas, P. (1974) Arch. Biochem. Biophys. 164:518-526.
42. Cohen, S., Fava, R.A., and Sawyer, S.T. (1982) Proc. Natl. Acad. Sci. USA 79:6237-6241.
43. Benveniste, R., Speeg, K.V., Jr., Carpenter, G., Cohen, S., Linder, J., and Rabinowitz, D. (1978) J. Clin. Endocrinol. Metab. 46:169-172.
44. Libermann, T.A., Neubaum, H.R., Razon, N., Kris, R., Lax, I., Soreq, H., Whittle, N., Waterfield, M.D., Ullrich, A., and Schlessinger, J. (1985) Nature 313:144-147.
45. Hendler, F.J., and Ozanne, B.W. (1984) J. Clin. Invest. 74:647-651.
46. Livneh, E., Glazer, L., Segal, D., Schlessinger, J., and Shilo, B.-Z. (1985) Cell 40:599-607.
47. Cohen, S., Ushiro, H., Stoscheck, C., and Chinkers, M. (1982) J. Biol. Chem. 257:1523-1531.
48. Carpenter, G., and Cohen, S. (1977) Biochem. Biophys. Res. Commun. 79:545-552.
49. Carpenter, G., King, L., Jr., and Cohen, S. (1979) J. Biol. Chem. 254:4884-4891.
50. Ushiro, H., and Cohen, S. (1980) J. Biol. Chem. 255:8363-8365.
51. Cohen, S., Carpenter, G., and King, L., Jr. (1980) J. Biol. Chem. 255:4834-484 .
52. Cohen, S., Ushiro, H., Stoscheck, C., and Chinkers, M. (1982) J. Biol. Chem. 257:1523-1531.
53. Buhrow, S.A., Cohen, S., and Staros, J.V. (1982) J. Biol. Chem. 257:4019-4022.
54. Buhrow, S.A., Cohen, S., Garbers, D.L., and Staros, J.V. (1983) J. Biol. Chem. 258:7824-7827.
55. Ullrich, A., Coussens, L., Hayflick, J.S., Dull, T.J., Gray, A., et al. (1984) Nature 309:418-425.
56. Carpenter, G. (1987) Ann. Rev. Biochem. 56 (in press).
57. Yamamoto, T., Nishida, T., Miyajima, N., Kawai, S., Ooi, T., et al. (1983) Cell 35:71-78.

58. Sawyer, S.T., and Cohen, S. (1981) Biochemistry 20:6280-6286.
59. Foulkes, J.G., Rosner, M.R. (1985) in <u>Molecular Mechanisms of Transmembrane Signalling</u> ed. P. Cohen, M.D. Houslay, p. 217-252. Amsterdam/New York: Elsevier.
60. Lin, C.R., Chen, W.S., Lazar, C.S., Carpenter, C.D., Gill, G.N., Evans, R.M., and Rosenfeld, M.G. (1986) Cell 44:839-848.

USE OF ANALOGS OF LH-RH AND SOMATOSTATIN FOR THE
TREATMENT OF HORMONE DEPENDENT CANCERS

Tommie W. Redding and Andrew V. Schally

Endocrine, Polypeptide and Cancer Institute,
VA Medical Center and Section of Experimental
Medicine, Department of Medicine, Tulane
University School of Medicine, New Orleans, LA
70146 U.S.A.

INTRODUCTION

Historically, the investigation of the relationship between hormone induced tissue growth and neoplasia began as early as 1786, when John Hunter recognized the physiological dependency of the prostate gland on testicular secretions (1). Later these findings were exploited for therapeutic purposes. White, in 1895, reported that patients with prostatic disease were significantly improved by surgical castration (2,3). About the same time (1896) the intuitive studies of Beatson (4) demonstrated a beneficial effect of ovariectomy in patients with breast cancer. However, the definitive link between modern oncology and endocrinology was not established until the pioneering work of Huggins et al (5,6) in the 1940's and the development of the concept that malignant prostate tumors are primarily an overgrowth of adult epithelial cells. They demonstrated that the elimination of androgens by either castration or the administration of estrogens inhibited prostatic cancer in man (5,6). This was the first time that rational hormonal means were utilized as a mode of treatment for prostate carcinoma. Other experimental studies by Loeb et al (7) and epidemiological studies by

Bittner and his associates have (8,9) led to recognition of hormonal, genetic and viral components in the development of breast cancer. It is now quite apparent that areas of contact between endocrinology and oncology are numerous, diverse and continually growing.

The discovery of hypothalamic hormones and the elucidation of their structures in the late 1960's and early 70's has led to a better understanding of the complex controls regulating the anterior pituitary gland and its target organs as well as offering new methods of hormonally manipulating endocrine organs (10,11). The development of superactive analogs of LH-RH and of somatostatin may open a new era in the treatment of some types of hormone sensitive cancers. This new, but radically different approach to endocrine therapy has already found clinical applications in the treatment of prostatic and other hormone-dependent cancers in men (10,11). The inhibition of growth of endocrine-dependent tumors by analogs of hypothalamic hormones was initially viewed as a sideline of the application of hypothalamic peptides but it now appears to be growing and of major clinical value. Ongoing investigations on the application of analogs of LH-releasing hormones and somatostatin in the field of neoplasia include treatment of breast and prostate cancer, pituitary tumors, chondrosarcomas and osteosarcomas, pancreatic cancer and other hormone-sensitive tumors. We will review and summarize the results of some of our basic and clinical investigations on the effects of agonistic and antagonistic LH-RH analogs and analogs of somatostatin on hormone-dependent tumors.

Prostate Cancer

The LH-RH agonists characterized by a D-amino acid in position 6 were originally developed to treat infertility (12,13). But it was soon discovered that while acute injections of these superactive analogs to males or females causes a marked and prolonged release of gonadotropins, paradoxically, their chronic administration results in atrophy of the gonads and accessory sex organs and a decrease in sex steroids

(14,15). The ability of these compounds to inhibit ovarian and testicular steroidogenesis and produce regression of steroid dependent reproductive structures led to the concept that sex steroid-responsive tumors might also be affected in a manner similar to that produced by gonadectomy or by the administration of antiandrogenic or antiestrogenic drugs.

The histological and biochemical similarities of the Dunning R-3327H prostate tumor to human prostate adenocarcinoma have made this tumor an acceptable model for the study of human prostate cancer. From 70% to 90% of the cells of this tumor appear to be dependent of androgen for maximal growth stimulation (16,17). Rats bearing this tumor respond for a long period of time to surgical castration by tumor involution. However, after this initial response to castration, the tumor relapses and growth is reinitiated (10,11). Thus this tumor was chosen for our studies since it mimics many of the properties of human prostrate carcinoma.

Treatment of male Copenhagen X Fisher, F-1 rats bearing the Dunning R-3327H prostate adenocarcinoma with D-Trp-6-LH-RH, 25 µg/day for 42 days, reduced the percentage increase in tumor volume to one-third and decreased the actual tumor weight by 58% as compared to untreated controls (18). Tumor doubling time was more than 4 times longer in rats receiving D-Trp-6-LH-RH than in controls. This treatment also diminished the weights of both the ventral prostate and testes, but had no effect on the weight of the anterior pituitary gland. Serum LH and FSH levels were significantly decreased in rats receiving this analog. Serum prolactin and testosterone levels were significantly reduced after treatment with D-Trp-6-LH-RH, whereas progesterone levels were increased. This was the first study to demonstrate the potential clinical efficacy of D-Trp-6-LH-RH in the treatment of prostate carcinoma and other hormone dependent tumors in man (18).

Development of Microcapsules of D-Trp-6-LH-RH

The use of D-Trp-6-LH-RH for the treatment of endocrine dependent tumors has been greatly enhanced by the ability to administer this peptide using a practical delivery system and a regimen capable of maintaining a controlled constant level of peptide over an extended period of time (19). The previous modes of administration of D-Trp-6-LH-RH were limited by the fact that the analog is initially delivered to the target tissue but thereafter its levels in blood continuously decline. The i.v., s.c. or intranasal modes of administration, with their fluctuating blood levels of the peptide, may not achieve the optimal pharmacological effects and may therefore necessitate higher dosage regimens and frequency of administration. Our analog D-Trp-6-LH-RH administered in a preprogrammed and unattended delivery system and at a constant rate and duration of time, provides a continuous and complete suppression of testosterone levels which greatly enhances its ease of use and efficacy.

A polymer of poly (d,l-lactide-co-glycolide) has been used by us to formulate microcapsules containing D-Trp-6-LH-RH (19). These microcapsules are biodegradable and biocompatible with living tissue. Photomicrographs, obtained by scanning electron microscopy, show that the microcapsules were 50 microns or less in diameter and smooth, indicating that a continuous coating of polymer was present. They are designed for a constant controlled release of the LH-RH analog over a 30 day period. Since there is minimal patient intervention, these microcapsules offer the additional advantage of higher patient compliance (19).

To monitor the release of D-Trp-6-LH-RH from microcapsules and to study the pharmacokinetics of liberation of D-Trp-6-LH-RH from the microcapsules we developed a sensitive and specific RIA for D-Trp-6-LH-RH in unextracted serum (20). Adult male Sprague-Dawley rats received single i.m. injections of microcapsules of D-Trp-6-LH-RH. This batch of microcapsules produced a protracted elevation of serum D-Trp-6-LH-RH levels over a period of at least 30 days. The profile of this release was biphasic in nature. The peak of phase 1 release was on day 5 and that of phase 2 on day 18. The release of the D-Trp-6-LH-RH from these microcapsules was distributed over a period of about 30 days. Serum RLH values in rats injected with microcapsules were highest on day 1, and declined to basal by day 5. Serum testosterone was elevated on day 1 but fell to levels below normal by day 3 and remained suppressed for 30-40 days. The suppression of serum testosterone demonstrates that the microcapsules exerted biological effects for at least 30 days. The depression of serum testosterone in the presence of substained LH levels has been observed previously and suggests that chronic administration of LH-RH agonist suppresses the gonads more than the pituitary, and/or that the immunologically active LH detected by RIA may have reduced biological activity (20).

In the first experiment Dunning tumor rats were injected intramuscularly with microcapsules designed to release D-Trp-6-LH-RH at a constant rate of about 25 µg over a 24 hour period for 30 days, while another group was given unencapsulated D-Trp-6-LH-RH subcutaneously at a dose of 12.5 µg twice a day (19). In rats injected with microcapsules tumor weights were decreased by 81% (p=0.05), whereas in the group injected daily s.c. there was only a 38% reduction in tumor weights as compared to controls.

Nonparametric analysis of the percentage change in individual responses of tumors indicated that D-Trp-6-LH-RH, given either as a s.c. injection or by intramuscular injection of microcapsules, significantly decreased tumor volumes. Chronic treatment with D-Trp-6-LH-RH by either route of administration also significantly decreased ventral prostate weights as compared to controls. Testes weights in the group given microcapsules and injected subcutaneously were reduced by 60% (p=0.01) and 42% (p=0.01) respectively, but neither group showed significant changes in body weight or the weights of the anterior pituitary gland or adrenal glands (19).

In the second experiment, using the same design as in the first, D-Trp-6-LH-RH was given s.c. at a dose of 25 µg bid, double that of the microcapsule formulation.

There was a reduction in tumor and gonadal weights as well as accessory sex organs similar to that seen in the first experiment (19). Doubling the s.c. dose of D-Trp-6-LH-RH was no more effective in decreasing tumor weight than the microcapsules. Again, on a percentage change basis, both treatments reduced tumor volume. Body weights, anterior pituitary and adrenal weights were not altered by this treatment as compared to control.

In both experiments, testosterone levels were suppressed by more than 90% by day 5 in animals injected with microcapsules as compared to control tumor rats. By day 19 testosterone fell to undetectable levels and was still undetectable by our RIA methods on day 30. In rats given D-Trp^6-LH-RH s.c., testosterone fell by more than 70% after 5 days of treatment but remained at detectable levels throughout the experiment. Administration of D-Trp-6-LH-RH either as microcapsules or s.c. injection significantly decreased serum prolactin levels but did not alter serum growth hormone levels (19).

Somatostatin Analogs

Modern superactive octapeptide analogs of somatostatin, including D-Phe-Cys-Tyr-D-Trp-Lys-Val-Cys-Thr-NH$_2$ (RC-121) in doses of 2.5 µg b.i.d., significantly decreased the weight and volume of Dunning R3327H prostate cancers and, when given in combination with once a month D-Trp-6-LH-RH microcapsules, potentiated the effects of the latter (21). Microcapsules of D-Phe-Cys-Tyr-D-Trp-Lys-Val-Cys-Trp-NH$_2$ (RC-160), designed for a controlled release of this analog for 30 days, also inhibited the growth of Dunning prostate tumors when given alone. Combination of microcapsules of D-Trp-6-LH-RH with microcapsules of RC-160 resulted in a synergistic potentiation of the inhibition of prostate tumors. The combination of LH-RH agonists with somatostatin analogs could result in an increase in the therapeutic response in patients with prostate cancer (21). Prolactin and growth hormone could be involved in prostate cancer as co-factors (10,11). Somatostatin analogs inhibit prolactin and growth hormone secretion and have direct antiproliferative effects on cells (22). The reduction

in prolactin and growth hormone levels produced by the
administration of somatostatin analogs, combined with the
decrease in serum testosterone resulting from chronic
treatment with LH-RH agonists, may inhibit growth of
prostate tumors better than LH-RH agonists alone (10,11).
Evidence obtained in animal models of prostate tumors is
in agreement with this hypothesis.

Human Studies

There is evidence also for the efficacy of periodic
administration of D-Trp-6-LH-RH microcapsules for
suppressing testicular secretion in patients with
prostate cancer (23). Patients received,
intramuscularly, an injection equivalent to 3mg
Decapeptyl in microcapsules designed to deliver, in a
controlled fashion, a daily dose of 100 g D-Trp-6-LH-RH
for 30 days. Seven men with a prostatic carcinoma
received subcutaneously 500µg Decapeptyl/day for seven
days and in addition received on days 8, 28, and 56 an
i.m. injection of Decapeptyl microcapsules. Plasma LH
and FSH peaked on day 2. In 5 men high testosterone
levels were maintained from day 2 to day 4, and then fell
abruptly on day 6. In the other two subjects,
testosterone secretion was completely suppressed after
the second i.m. injection.

The safety and efficacy of a delayed release
formulation of D-Trp-6-LH-RH microcapsules was compared
with orchiectomy in the treatment of advanced prostatic
microcapsules and 38 patients to orchiectomy.
Suppression of testosterone and reduction in prostatic
acid phosphatase were similar in both groups. Overall,
87% of the patients in the D-Trp-6-LH-RH group and 81% of
the orchiectomy group had an objective response to
treatment. Side effects related to the decrease in
testosterone were similar in both groups. Three patients
in the D-Trp-6-LH-RH group experienced a disease "flare"
in the first 10 days of treatment, which resolved
completely with the fall in testosterone to castrate
levels. There was a trend on follow-up towards decreased
psychological morbidity in the D-Trp-6-LH-RH group. It
was concluded that the slow-release preparation of

D-Trp-6-LH-RH microcapsule is equally efficacious as orchiectomy and offers an important new method free of side effects in the management of advanced prostatic carcinoma.

Effect of Combination of LH-RH Agonists with Antiandrogen

During the past four decades, palliative treatment for prostate cancer has been largely based on the elimination of androgens. Castration and estrogen administration have been routinely employed as primary therapies. More modern methods of reducing androgenic stimulation of prostate cancer include inhibitors of androgen synthesis and antiandrogens (25,26,27). Administration of aminoglutethimide inhibits adrenal steroidogenesis, including that of androgens, and has been used as an adjunct therapy with orchiectomy in the treatment of prostatic cancer (26,27). The term antiandrogen has been used for substances which interfere with the action of androgens.

Several reports by one group (28,29) claimed that administration of nonsteroidal antiandrogens RU 23908 (Anandron) or flutamide, in combination with daily injections of LH-RH agonist Buserelin was more effective clinically in the treatment of prostatic carcinoma than the use of the LH-RH agonist alone. This prompted us to investigate and compare the effect of the antiandrogen flutamide and long acting microcapsules of D-Trp-6-LH-RH given alone or in combination on the growth of the androgen-dependent Dunning R-3327H prostate tumor in rats (30).

Groups of rats were injected s.c. with flutamide at a dose of 25 mg/kg or about 9.3 mg/day or given i.m. injections of microcapsules designed to release D-Trp-6-LH-RH at a rate of about 25 µg/day. Another group was treated with a combination of daily injections of flutamide and a once a month injection of D-Trp-6-LH-RH microcapsules. The treatment with these drugs was continued for 60 days. Tumor growth expressed as percentage change from initial volume in all three experimental groups was inhibited within 7 days after treatment began. After 60 days, tumor volume was significantly reduced in the group treated with microcapsules (281 ± 49%) as compared to control rats (2852 ± 1124%; p<0.01). Flutamide alone also significantly decreased tumor volume to 1054 ± 371% (p<0.05). The microcapsules given together with flutamide similarly reduced tumor volume to 519 ± 138% (p<0.01), but no evidence of a synergistic effect of this combination was found.

Tumor weights were significantly decreased by about 43% in rats treated with flutamide as compared to control rats (6.84 g vs. 12.08 g, p<.01). Testes weights were not changed, but the ventral prostate weight was significantly reduced. Treatment of rats bearing prostate tumors with D-Trp-6-LH-RH microcapsules resulted in a greater decrease (73%) in tumor weight (3.22 g, p<.01 vs. control) than in the case of flutamide, although this difference was not statistically significant. The combined administration of D-Trp-6-LH-RH with flutamide also significantly decreased tumor weight to 4.26 g, (p<.01) but this reduction was smaller than that produced by D-Trp-6-LH-RH microcapsules alone. Within 7 days after the administration of flutamide, serum testosterone levels rose significantly, reaching values 10-11 times greater than the controls by day 30 and remained significantly elevated throughout the 60 days of treatment with this antiandrogen. In rats given D-Trp-6-LH-RH microcapsules, testosterone levels fell by 95% within 7 days as compared to controls and remained at non-detectable levels throughout the remainder of the experiment. The treatment with microcapsules of D-Trp-6-LH-RH combined with flutamide decreased testosterone levels by more than 99% by day 7 and resulted in non-detectable values between days 14 to 60.

These findings are in excellent agreement with several previous studies with microcapsules of D-Trp-6-LH-RH (10,11,19). Flutamide did not have a synergistic effect on tumor inhibition when administered together with microcapsules of D-Trp-6-LH-RH. In fact, the combination of these two drugs appeared to suppress tumor growth less than microcapsules of D-Trp-6-LH-RH given alone. From our experimental data it is doubtful that antiandrogens, at least not flutamide, can increase the efficacy of long-acting microcapsules of D-Trp-6-LH-RH.

While this data in rats may not be extrapolated unequivocally to men with advanced prostate carcinoma, they nevertheless leave unanswered the question whether the combination of LH-RH analogs with antiandrogens (28,29,31) is more effective in the latter species than LH-RH analogs alone, since the results of therapeutic approaches in rats with Dunning tumors usually correspond to what happens in men with prostate carcinoma (18).

Combination of Hormonal Therapy with Chemotherapy

It is becoming increasingly apparent that carcinomas of the prostate are heterogeneous and contain both hormone-dependent and independent cells (17,18,32). After androgen ablation, most of the hormone dependent cells stop proliferating and die (32,33), but hormone independent cells, by a clonal selection process, eventually repopulate the tumor. As the prostate cancer progresses to a hormone independent state, this produces a relapse to androgen ablation therapy (17,18,32,33). In another series of studies, we investigated a possible synergistic effect of combining a modern hormonal therapy with the administration of the cytotoxic drugs cyclophosphamide or mitoxantrone (Novantrone) in rats bearing hormone-dependent Dunning R-3327H prostatic adenocarcinoma at two stages of development (34,35).

D-Trp-6-LH-RH and Cytoxan

In the first experiment, treatment with D-Trp-6-LH-RH microcapsules and/or cytoxan was started 90 days after the transplantation when the Dunning R-3327H prostate tumors were well developed, having a volume of about 3000 mm^3. One group of rats was injected intramuscularly once a month with microcapsules designed to release D-Trp-6-LH-RH at a controlled rate for 30 days. Another group of rats received i.p. injections of cyclophosphamide twice a week. The third group of rats was treated with a combination of D-Trp-6-LH-RH microcapsules and cyclophosphamide. The treatment with these drugs was continued for 60 days. Tumor growth in all three experimental groups, as measured by percentage change from the initial tumor volume, was inhibited within 15 days after treatment was begun. After 60 days, tumor volume was significantly reduced in the group treated with microcapsules (474 ± 133%; $p<0.05$) as compared to untreated control rats (1882 ± 304%). Cyclophosphamide alone also decreased tumor volume to 735 ± 154% ($p<0.05$). The microcapsules given together with cytoxan similarly reduced tumor volume to 523 ± 115% ($p<0.05$), but no synergistic effects of this combination was found (34).

In a second experiment, rats with early developing tumors, with a tumor volume of 60-70 mm^3, were used. The same design as in the first experiment was followed, but the treatment period was extended to 90 days. The survival rate was virtually 100%. Early treatment of rats bearing prostate tumors with cyclophosphamide, D-Trp-6-LH-RH microcapsules or the combination of both significantly inhibited tumor growth throughout the experimental period. The microcapsules were more effective than cytoxan and the combination of the two drugs appeared to completely arrest tumor growth. After 90 days of treatment, tumor volumes were significantly reduced in all three experimental groups as compared to control rats which had a mean tumor volume of 17543 ± 2996 mm^3. In rats treated with the combination of cyclophosphamide and D-Trp-6-LH-RH microcapsules, tumor volume (31 ± 10 mm^3; $p<0.01$ vs. controls) was smaller than in the groups injected only with microcapsules (883 ± 406 mm^3; $p<0.01$) or with cyclophosphamide alone (3018 ± 787, $p<0.01$). These conclusions based on the tumor volume were confirmed by determination of tumor weights on autopsy. Tumor weights were significantly reduced in all the experimental groups, but the decrease was smaller in rats receiving cyclophosphamide than in the group injected with D-Trp-6-LH-RH microcapsules. The combined administration of cyclophosphamide with D-Trp-6-LH-RH microcapsules resulted in tumor weights which were about

10 times smaller than those in rats treated with microcapsules alone.

Our studies showed that when the treatment of rats was started at the time that prostate tumors were well developed and continued for two months, hormonal therapy in the form of microcapsules and cytoxan given alone caused some inhibition of tumor growth, but none of the groups exhibited complete tumor involution, and there was no evidence of a synergistic effect. On the other hand, if the therapy was initiated early in the progression of tumor growth and continued for 90 days, the individual therapeutic efficacy of either treatment was greatly enhanced and the microcapsules reduced tumor volume and weight much more effectively than cytoxan. This could indicate a powerful suppression of the growth of predominantly hormone-dependent tumor cells. Under these conditions, the combination of microcapsules plus cytoxan virtually arrested tumor growth.

D-Trp-6-LH-RH Microcapsules and Novantrone

In another study we used Novantrone (mitoxantrone), which is a modern synthetic chemotherapeutic agent with a mode of action similar to that of the anthracycline antibiotic, Adriamycin, but which is less toxic (35). Novantrone was selected on the basis of its property to intercalate into DNA, resulting in blocking DNA and RNA synthesis in tumor cells, and its activity in a wide range of experimental animal tumors. Novantrone can exert inhibitory effects on both proliferating and non-proliferating cells (35,36). The aim of this investigation was to determine whether the combination of hormonal therapy using microcapsules of D-Trp-6-LH-RH and chemotherapy based on Novantrone would enhance the inhibition of tumor growth in the Dunning R3327H prostate cancer model. Tumor volume was significantly reduced by the administration of microcapsules (966 ± 219 mm^3) ($p<0.01$) or Novantrone (3606 ± 785 mm^3) ($p<0.01$) as compared to the control group ($14,476 \pm 3,045$ mm^3) ($p<0.01$). Throughout the treatment period of 70 days, the microcapsules of D-Trp-6-LH-RH reduced tumor volume more than Novantrone given alone. However, the

combination of these two agents caused a greater
inhibition of tumor growth (189 ± 31 mm^3) than the
microcapsules or Novantrone administered alone (p<0.01
vs. control; p<0.005 vs. microcapsules by Student's
t-test).

The effects of the administration of the two agents
on the percent change in tumor volume during the period
of the experiment was also followed. Using this
parameter, the inhibition induced by the D-Trp-6-LH-RH
microcapsules (672 ± 153% compared to the control of
10,527 ± 1,803%, p<0.01 vs. control) was again much
greater than that of Novantrone (2722 ± 421% p<0.01 vs.
control) when given alone. The combination of the two
agents (105 ± 29%) (p< 0.01 vs. control) was again
clearly the most effective and resulted in the greatest
inhibition of the percent increase in tumor volume (35).
This inhibition was significantly greater than that
obtained with microcapsules alone (p<0.005 by Student's
t-test). These conclusions based on tumor volume were
confirmed by determination of tumor weights on autopsy.

Histologically, all the tumors were
well-differentiated adenocarcinomas (37). The tumors in
the control group presented a glandular pattern and the
glands were lined by columnar epithelial cells. Some
mitoses were present and homogenous eosinophilic and PAS
positive material was found in the lumens of the glands.
The glands were supported by a delicate and scant stroma.
The tumors in the group treated with microcapsules of
D-Trp-6-LH-RH alone, also presented a glandular pattern,
but the epithelial cells were cuboidal or flat and
sometimes formed cysts instead of glands. The number of
these cells was decreased as compared with the control
group and the interglandular tissue was increased by a
prominent proliferation of collagenous fibrils,
fibroblasts, histiocytes and myoepithelial cells. The
tumors in the group treated with Novantrone alone showed
a decreased number of epithelial cells as compared to the
control group, but the cells were more numerous than in
the groups treated with microcapsules of D-Trp-6-LH-RH.
Tumors treated with the combination of microcapsules and
Novantrone showed the largest decrease in the number of
epithelial cells and the greatest increase in the amount

of connective tissue of all the groups. Connective tissue fibers surrounded individual tumoral cells and several layers of the collagenous fibers were observed among interstitial cells. The tumoral cells were atrophic, either cuboidal or flat in shape.

These histological results suggest that all three therapies inhibit the growth of the tumoral cells, and these cells are replaced by connective tissue. This effect is more pronounced in the group treated with the combination of microcapsules and Novantrone (37).

Although our experimental conditions do not exactly duplicate the clinical picture, our findings appear to reinforce the view put forward on the basis of work in experimental models of prostate cancer (10,11,34,35) and clinical observations (23,24) that the combined therapy should be initiated soon after diagnosis is established. The efficacy of the combination of LH-RH agonist microcapsules with various chemotherapeutic agents remains to be established in men with prostatic carcinoma, but an improvement in the therapeutic response and an increase in the survival rate appear as attractive targets for such investigations.

Chondrosarcomas and Osteosarcomas

It has been established that the Swarm chondrosarcoma, as in the case of normal cartilage, is hormone-responsive and its growth depends on growth hormone, somatomedin and glucocorticoids (38,39). Chronic administration of D-Trp-6-LH-RH to rats bearing transplantable Swarm chondrosarcoma at a dose of 30 µg/day for 30 days caused a significant reduction in tumor weight (38). When D-Trp-6-LH-RH was administered together with pNH_2-Phe^4-SS, there was a similar decrease in tumor weight, which was highly significant ($p<.01$). In the second experiment D-Trp-6-LH-RH alone or in combination with D-5F-Trp^8-SS decreased tumor volume after 14 days of treatment ($p<.05$), although tumor weights were not significantly altered (38).

In another experiment, D-Trp-6-LH-RH, at a dose of 30 μg b.i.d. for 22 days, diminished tumor weight and volume without affecting anterior pituitary and adrenal weights. Both the testes and ventral prostate weights were significantly decreased. Serum levels of GH and prolactin in rats bearing the Swarm chondrosarcoma were significantly reduced in rats treated with D-Trp-6-LH-RH. There appeared to be a greater suppression of serum GH and prolactin levels when D-5F-Trp^8-SS was administered together with D-Trp-6-LH-RH (38).

While the inhibition of the growth of chondrosarcomas by analogs of somatostatin could be explained by their suppressive effects on circulating GH and insulin levels, the mechanism of the suppression of these tumors by D-Trp-6-LH-RH is not clear. It has been suggested that chronic administration of D-Trp-6-LH-RH may suppress the pituitary-adrenal axis or the adrenal itself. Chronic treatment with D-Trp-6-LH-RH also increases blood progesterone levels in normal rats and in rats bearing the Dunning R-3327 prostate tumors (18). Salomon et al. (39) found high affinity glucocorticoid binding sites in the cytosol fraction of the Swarm chondrosarcoma and reported that dexamethasone and progesterone were active competitors for these binding sites. It is possible that increased progesterone levels following chronic treatment with D-Trp-6-LH-RH interfere with the growth and metabolism of the chondrosarcoma through a receptor-mediated pathway.

The inhibition of growth of the Swarm chondrosarcoma in rats by analogs of hypothalamic hormones (38) raises some hopes that eventually such analogs might lead to a new endocrine therapy for chondrosarcoma, osteosarcomas and related hormone-dependent neoplasias.

Breast Cancer

Various experimental studies suggest that analogs of LH-RH might be useful for treatment of estrogen-dependent breast cancer (40,41,42,43). Tumor weights in BDF1 mice bearing the MXT 3.2 mammary adenocarcinoma were significantly decreased after administration of 25 μg/day of D-Trp-6-LH-RH for 21 days (44). The final tumor volume was also reduced by 47%. Ovarian weights were decreased by only 22% but final body weights were not changed from those of controls. Treatment with D-Trp-6-LH-RH lowered plasma progesterone, but not to ovariectomy levels (44). Since the mice are reputedly refractory to inhibitory effects of LH-RH agonists (45), it is possible that D-Trp-6-LH-RH induced mammary tumor regression by some other mechanism.

In Wistar-Furth rats bearing the MT/W9A mammary adenocarcinoma, treatment with 25 µg of D-Trp-6-LH-RH b.i.d. for 28 days significantly decreased tumor weights by more than 50% and tumor volume by 67% (44). The percentage change in tumor volume (-51%), based on individual responses, was also highly significant when compared to controls. Ovarian weights fell by 38% and anterior pituitary weights were reduced by 21%, but the final body weights remained unchanged. A small number of rats were ovariectomized after transplantation of the tumor to test for the estrogen dependency of the carcinoma. After 28 days tumor weights were decreased by more than 90% and in some rats the tumor had disappeared.

The effects of two powerful antagonistic analogs of LH-RH, Ac-D-p-Cl-Phe^{1,2}, Phe^3, D-Arg^6, D-Ala^{10}-LH-RH and N-Ac-p-Cl-Phe^{1,2}, D-Trp^3, D-Arg^6, D-Ala^{10}-LH-RH for 3 weeks significantly reduced the weight and volume of estrogen dependent MXT 3.2 mammary tumor in BDF1 mice. Ovarian weights and plasma levels of progesterone were diminished after treatment with this antagonist. N-Ac-D-p-Cl-Phe^{1,2}, D-Trp^3, D-Arg^6, D-Ala^{10}-LH-RH, administered for 4 weeks in doses of 50 µg/day to Wistar-Furth rats reduced the weight by 58% and the volume of MT/W9A mammary tumor by 42%. The percentage of change in tumor volume (-40%) was similarly highly significant after treatment with the antagonist. LH, estrogen and progesterone levels were also greatly reduced in rats treated with the antagonist. The suppression of tumor growth by the antagonists is most likely linked with the inhibition of the levels of sex steroids, but some direct of the antagonists on mammary tumors cannot be excluded.

Other studies in Wistar/Furth rats bearing estrogen-and prolactin-dependent MT/W9A mammary adenocarcinoma showed that once-a-month administration of microcapsules releasing 25 µg/day of D-Trp-6-LH-RH, or twice daily injections of 3 µg somatostatin analog (RC-15) Ac-p-Cl-D-Phe-Cys-Phe-D-Trp-Lys-Thr-Cys-Trp-NH$_2$ (RC-160) or 25 µg D-Trp-6-LH-RH per day significantly inhibited the growth of MT/W9A mammary tumor (46). When both types of microcapsules were given together, a significant synergism between the LH-RH agonists and the somatostatin analog was demonstrated in the inhibition of tumor growth. It appears that the reduction in prolactin and GH levels induced by administration of a somatostatin analog, combined with the decrease in estrogen values which results from chronic treatment with LH-RH agonists, leads to a greater inhibition of mammary tumors than that obtained with LH-RH agonists alone. It is also possible that somatostatin analogs could also exert direct antiproliferative effects on breast cancer cells by inhibiting autocrine growth factors (47).

Regression of mammary tumors in rats and mice in response to analogs of LH-RH suggested that these compounds should be considered for a new hormonal therapy for breast cancer in women. So far, only limited clinical trials have been carried out, but regression of tumor mass and disappearance of metastases in premenopausal and postmenopausal women with breast cancer treated with D-Trp-6-LH-RH, Buserelin or Leuprolide have been reported (10,11,48,49). The use of microcapsules makes the treatment of breast cancer with LH-RH agonists more convenient and ensures patient compliance. It is also possible that treatment with LH-RH agonists can be combined with chemotherapy or with peptides that inhibit prolactin and growth hormone release such as somatostatin analogs or PIF. Since prolactin may act as a promoter in the development or growth of mammary tumors (10,11) and GH stimulates the production of the growth factor somatomedin, the reduction in prolactin and growth hormone levels induced by somatostatin analogs and/or PIF may potentiate inhibition of mammary tumors obtained with LH-RH analogs alone (44). In animal studies a significant synergism between the LH-RH agonist and the somatostatin analog in the inhibition of tumor growth was demonstrated when both peptides were given together.

Pancreatic Cancer

Regulation of growth of the exocrine pancreas appears to be hormonal in nature. The gastrointestinal hormones, cholecystokinin (CCK) and secretin are the chief endocrine stimulants of pancreatic exocrine secretions (50). An important, recently established action of CCK, secretin and gastrin is their ability to stimulate the growth of the exocrine pancreas (50,51,52,53). The role which these gastrointestinal hormones play in the development and growth of pancreatic cancer is not clearly understood, but it is likely that they may influence the growth of malignant cells of the pancreas (54).

The inhibition of release of gastrointestinal hormones by administration of somatostatin or its various analogs is well documented (55). Somatostatin inhibits not only the liberation of insulin and glucagon, but also the release and/or action of secretin, gastrin, CCK and other gastrointestinal peptides such as VIP and motilin (56,57,58,59). Somatostatin has been shown to suppress exocrine pancreatic secretion in rats, dogs and humans (56,57,59,60).

Sex steroids may also play a role in the growth of normal and cancerous pancreas (61,62,63). The presence of specific receptors for estrogen and androgen in pancreatic cells indicates that sex hormones may influence neoplastic cell processes (64,65). All these findings suggest that pancreatic adenocarcinomas may be sensitive to both gastrointestinal and sex hormones.

Using animal models of pancreatic cancer, we investigated the effect of analogs of hypothalamic hormones on the growth of pancreatic tumors (66). In Wistar/Lewis rats bearing the acinar pancreatic tumor DNCP-322, chronic administration of L-5Br-Trp-8-Somatostatin significantly decreased tumor weights and volume. D-Trp-6-LH-RH also decreased tumor weight and volume. In Syrian hamsters with a ductal form of pancreatic cancer, administration of L-5-Br-Trp-8-S-S for 21-30 days, diminished tumor weights and volume. D-Trp-6-LH-RH, given twice daily or injected in the form of constant-release microcapsules, significantly decreased tumor weight and volume and suppressed serum levels of testosterone (66).

Other experiments were perfomed in hamsters with N-nitrosobis (2-oxypropyl) amine (B.O.P.)-induced ductal pancreatic cancers (67). Fifty female hamsters received B.O.P. 10 mg/kg every week for 18 weeks, then subsequently divided into treatment groups. The controls received no treatment. One group was treated with once-a-month microcapsules of D-Trp-6-LH-RH releasing 25 µg/day and another group was treated with microcapsules of somatostatin analog RC-160 releasing 5 µg/day. The treatment was continued for 45 days. The survival rate was 35% for controls, 70% for D-Trp-6-LH-RH treated group and 70% for the groups that received the somatostatin analog. Tumor volumes in the D-Trp-6-LH-RH treated group were 0.33 ± 0.12 cm^3 and in the RC-160-treated group, 0.65 ± 0.22 cm^3 vs. 1.36 ± 0.3 cm^3 for controls. Tumor weights were 0.88 ± 0.12g for controls, 0.2 ± 0.07 g for D-Trp-6-LH-RH and $0.43 \pm .17$ g for the RC-160 treated group. These results in the female animals with induced ductal tumors confirm and extend our findings that D-Trp6-LH-RH and somatostatin analogs inhibit the growth of pancreatic tumor.

D-Trp-6-LH-RH decreases the growth of pancreatic carcinomas by probably creating a state of sex hormone deprivation (66,67). Somatostatin analogs reduce the growth of pancreatic ductal and acinar cancers, probably by inhibiting the release and/or stimulatory action of gastrointestinal hormones and other growth factors on tumor cells (66,67). It is also possible that some somatostatin analogs act directly on tumor tissue since somatostatin-14 has been shown to have antiproliferative effects on cells and to nullify the growth stimulation produced by EGF (68).

CONCLUSIONS:

A new approach to the treatment of various endocrine dependent tumors based on LH-RH analogs, superactive analogs of somatostatin and other analogs of hypothalamic hormones is being developed and appears to be promising and important because it is essentially devoid of side effects. Therapeutic regimens based on the use of microcapsules of LH-RH analogs and somatostatin analogs might supplement, or in some cases, replace, conventional procedures for the treatment of hormone-sensitive cancers. Experimental studies showed that agonists of LH-RH might be useful for treatment of androgen-dependent prostate cancer and estrogen-dependent breast cancer. Clinical results accumulated during the past few years support this view. On the basis of experimental findings, LH-RH agonists could also be considered as adjuncts for therapy of chondrosarcomas, osteosarcomas and pancreatic cancer, but the prinicpal approach for the treatment of these neoplasms might be based on somatostatin analogs. In turn, somatostatin analogs could serve as adjuncts in treatment of prostate cancer and breast cancer. Long-acting delivery systems for once-a-month administration of microcapsules of D-Trp-6-LH-RH have been successfully combined with some chemotherapeutic agents such as Cytoxan and Novantrone in the experimental treatment of prostate cancer. Work is in progress on the use of LH-RH analogs for the treatment of ovarian cancer and neoplasms of the female genital tract.

ACKNOWLEDGEMENTS

The experimental work described in this paper was supported by National Institutes of Health Grants AM07467, N.C.I Grants, CA 40003, CA 40077 and CA 40004 and by the Medical Research Service of the Veterans Administration.

REFERENCES

1. Hunter, J. (1786) "Observations in Certain Parts of the Animal Occonomy", 1st Edition London.
2. White, J.W. (1893) Amer. Surg. 18, 153.

3. White, J.W. (1895) Ann. Surg. 22, 1.
4. Beatson, G.T. (1896) Lancet ii:104
5. Huggins, C. and Hodges, C.V. (1941) Cancer Res 1:293.
6. Huggins, C. (1967) Science 156:1050-1054.
7. Lach, L (1940) J. Natl. Cancer Inst. 1:169-195.
8. Bittner, J.J. (1946-1947) Harvey Lect. 42:221.
9. Bittner, J.J. (1952) Texas Rep. Biol. Med. 10:160-246.
10. Schally, A.V., Comaru-Schally, A.M. and Redding, T.W. (1984) Proc. Soc. Exptl. Biol. Med. 175:259-281.
11. Schally, A.V., Redding, T.W. and Comaru-Schally, A.M. (1984) Med. Oncol. Tumor Pharmacothes. 1:109-118.
12. Schally, A.V., Coy, D.H. and Arimura, A. (1980) Int. J. Gynaecal. Obstet. 18:318-324.
13. Schally, A.V., Arimura, A. and Coy, D.H. (1980) Vitam. Horm. 38:257-323.
14. Johnson, E.S., Gendrick, R.L. and White, W.F. (1976) Fertil. Steril 27:853-860.
15. Rivier, C., Rivier, J. and Vale, W. (1979) Endocrinology 105:1191-1201.
16. Lesser, B. Bruchovsky, N. (1973) Biochem. Biophys. Acta 308:425-437.
17. Isaacs, J.I. (1984) The Prostate 5:545-557.
18. Redding, T.W. and Schally, A.V. (1981) Proc. Natl. Acad. Sci. USA 78:6509-6512.
19. Redding, T.W., Schally, A.V., Tice, T.R. and Meyers W.E. (1984) Proc. Natl. Acad. Sci. USA 81:5845-5848.
20. Mason-Garcia, M., Vigh, S. Comaru-Schally, A.M., Redding T.W., Somogyvari-Vigh, A., Horvath, J. and Schally, A.V. (1985) Proc. Natl. Acad. Sci. USA. 82:1547-1551.
21. Schally AV, Redding TW, Paz-Bouza JI, Bajusz S, Cai R-Z and Comaru-Schally AM Use of Analogs of Hypothalamic Hormones for the Treatment of Hormone-Dependent Cancers, Book Chapter in: Proceedings of the 14th International Cancer Congress, Budapest, August 1986 S. Karger Basle
22. Mascardo, R.N. and Sherline, P. (1982) Endocrinology 111:1394-1396.
23. Roger, M., Duchier, J., Lahlow, N., Nahoul, K. and Schally, A.V. (1985) Prostate 7:271-282.

24. Parmar, H., Lightman, S.L., Allen, L., Phillips, R.H., Edwards, L. and Schally, A.V. (1985).
25. Geller, J. and Albert, J.D. (1983) Semen. Oncol 10:(suppl 4)34-41.
26. Sanford, E.J., Argo, J.R., Rohner, I.J., Jr., Santen, R. and Lipton, A. (1976) J. Urol. 115:170-174.
27. Drago, J.R., Santen, R.S., Lipton, A., Worgul, T.J., Harvey, H.A., Boucha, A., Manni, A. and Rohna, I.J. (1984) Cancer 53:1447-1450.
28. Labrie, F., Dupont, A., Belanger, A., Cusan, L., Lacoursiere, Y., Monfette, G., Laberge, J.G., Emond, J.P., Fazekas, A.J.A., Raynaud, J.P., Husson, J.M. (1982) J. Clin. Invest. Med. 5:289-275, 1982.
29. Labrie, F., Dupon, A., Belanger, A., Emond, J. and Monfette, G. (1984) Proc. Natl. Acad. Sci. USA 81:2861-3863.
30. Redding, T.W., Schally, T.W. (1985) The Prostate 6:219-232.
31. Labrie,F., Belanger, A., Dupont, A., Emond, J., Lacoursiere, Y., and Monfette, G.
32. Isaacs, J.I. and Coffey, D.S. (1981) Cancer Res 41:5070-5075.
33. Lee, C., Jesek, C., Uke, E., Falkowski, W. and Gasyhack, I. (1981) in Hormone Related Tumors, eds Nagasawa, H. and Ahi, K., (Springer-Verlag, Berlin) pp. 261-285.
34. Schally, A.V. and Redding, T.W. (1985) Proc. Natl. Acad. Sci. USA 82:2498-2502.
35. Schally, A.V., Kook, A.I., Monje, E., Redding, T.W., Paz-Bouza, J.I. (1986) Proc. Natl. Acad. Sci. USA 83:8164-8768.
36. Neidhart, J.A., Gochmoar, D., Roach, R.W., Young, D. and Steinberg, J.A. (1984) Semin. Oncol. 11:11-31.
37. Paz-Bouza, J.I., Schar, N.A., Monje, E., Redding, T.W. and Schally, A.V. (1986) The Prostate
38. Redding, T.W. and Schally A.V. (1983) Proc Natl. Acad. Sci. USA 80:1078-1982.
39. Salomon, D.S., Paglia, L.M. and Verbruggen, L. (1979) Cancer Res. 39:4387-4395.
40. Nicholson, R.I. and Maynard, P.V. (1979) Br. J. Cancer 39:268-273.
41. Rose, D.P. and Pruitt, B. (1979) Cancer Res. 39:3968-3970.

42. De Sombre, E.R., Johnson, E.S. and White W.F. (1976) Cancer Res 36:3830-3833.
43. Dangry, A., Legros, N., Henson-Stienmon, J.A., Pastaels, J.L., Atassi, G. and Henson, J.C. (1977) Proc. Am. Assoc. Cancer Res. 20:87.
44. Redding, T.W. and Schally, A.V. (1983) Proc. Natl. Acad. Sci. USA 80:1459-1462.
45. Bex, F.G., Corbin, A. and France, E. (1982) Life Sci. 30:1263-1269.
46. Schally, A.V., Cai, R.Z., Torres-Aleman, I., Redding T.W., Szoke, B., Fu, D., Hierowski, M.T., Calaluca, J. and Konturck, S. (1986): In: Neural and Endocrine Peptides and Receptors, eds T.W. Moody (in press) Plenum, New York.
47. Redding, T.W. and Schally, A.V. in preparation.
48. Harney, H.A., Lipton, A. and Max, D. LH-RH analogs for human mammary carcinoma. In "Vickery, B.H., Nestor, J.J., Jr., Hafez ESE, eds. LH-RH and It's Analogs - Contraceptives and Therapeutic Applications. Lancaster, MTP Press, in press 1983.
49. Klijn, J.G.M. and DeJong, F.H. (1982) Lancet i:1213-1216.
50. Johnson, L.R. (1977) Gastroentesology 72:788-792.
51. Johnson, L.R. (1981) Cancer 47:1640-1645.
52. Dembinski, A.B. and Johnson, L.R. (1979) Endocrinology 105:769-773.
53. Dembinski, A.B. and Johnson, L.R. (1980) Endocrinology 106:323-328.
54. Townsend, C.M., Franklin, R.B., Watson, L.C., Glass, E.J. and Thompson, J.C. (1981) Surg. Forum 32:228-229.
55. Konturek, S.J., Cieszkowski, M., Bilski, J., Konturek, J., Bielanski, W. and Schally, A.V. (1985) Proc Soc Exp Biol Med 178(1):68-72.
56. Schally, A.V., Coy, D.H. and Meyer, C.A. (1978) Annu. Rev. Biochem 47:89-128.
57. Boden, G., Sivitz, M.C., Owen, O.C., Essa-Koumer, N. and Landor, J.H. (1975) Science 190:163-165.
58. Konturek, S.J. and Tasler, J., Cieszkowski, M., Coy, D.H. and Schally, A.V. (1976) Gastroenterology 70:737-741.
59. Konturek, S.J. and Tasler, J., Cieszkowski, M., Coy, D.H. and Schally, A.V. (1976) J. Clin Invest. 5811-6.

60. Adrian, T.E., Barnes, A.J., Long, R.G., O'Shaughnessy, D.J., Brown, M.R., Rivier, J., Vale, W., Blockburn, A.M. and Bloom, S.R. (1981) J. Clin. Endocrinol. Metab. 53:675-681.
61. Sandberg, A.A. and Rosenthal, H.E. (1979) J. Steroid. Biochem. 11:293-299.
62. Pousette, A. (1976) Bichem. J. 157:229-232.
63. Greenway, B., Igbal, M.J., Johnson, P.J. and Williams, R. (1981) Br. Med. J. 283:751-753.
64. Igbal, M.J., Greenway, B., Wilkinson, M.L., Johnson, P.J. and Williams, R. (1983) Clin. Sci. 65:71-75.
65. Greenway, B., Igbal, M.J., Johnson, P.J. and Williams, R. (1983) Br. Med. J. 286:93-95.
66. Redding, T.W. and Schally, A.V. (1984) Proc. Natl. Acad. Sci. USA 81:248-252.
67. Paz-Bouza, J.I., Redding, T.W. and Schally, A.V. (1986) Proc. Natl. Acad. Sci. USA (in press).
68. Hierowski, M.T., Liebow, C., des Sapin, K. and Schally, A.V. (1985) FEBS 179:252-256.

MULTI-HORMONAL REGULATION OF TYROSINASE EXPRESSION IN B16/C3 MELANOMA CELLS IN CULTURE

Terry J. Smith and Ellis L. Kline

Department of Medicine, State University of New York at Buffalo and the Veterans Administration Medical Center, Buffalo, NY 14215; and Department of Microbiology, Clemson University, Clemson, SC 29634

INTRODUCTION

Malignancy of the melanocyte, melanoma, accounts for roughly 2-3 percent of all skin neoplasms, but a majority of the mortality associated with them. Melanomas are difficult to treat; both surgical and chemotherapeutic maneuvers have failed to prove satisfactory. The natural course is extremely variable, with a period of months between diagnosis and death in some cases, and a long, protracted illness in others. They can spread by local invasion, by way of the lymphatics or by the hematogenous route and advanced disease is truly systemic.

Because the behavior of melanoma is unpredictable, an intensive search for factors which might influence tumor growth and metastatic spread has been undertaken. A number of investigators have examined the biological impact of a wide variety of humoral substances on these neoplasms in rodent models as well as in humans. Perhaps the most extensively studied and best characterized regulator of melanoma behavior is melanocyte-stimulating hormone (MSH), which has been shown to stimulate melanogenesis while inhibiting cellular proliferation (Wong and Pawelek, 1973). This response is associated with an elevated intracellular cAMP content (Pawelek, et al., 1973). The hormone binds to determinates on the outer cell membrane (Varga, et al., 1974). These are glycoproteins (Varga, et al., 1975) and probably represent functional receptors involved in the mediation of the response (Pomerantz and Chuang, 1970).

Disagreement between workers in the field exists concerning the mechanism involved in the stimulation of tyrosinase activity by MSH, however. Tyrosinase (EC 1:14.18.1), the rate limiting enzyme of melanin synthesis, catalyzes the conversion of tyrosine to dopa and dopa to dopa quinone (Pawelek, 1976), and is expressed in high abundance in melanocytes and in most melanomas. Evidence from one laboratory supports the activation of pre-existing enzyme molecules (Wong and Pawelek, 1975) in the Cloudman S91 cell line and contradicts the findings of Fuller and Viskochil (1979) who found that MSH stimulation in the same cell line could be completely blocked by either cycloheximide or actinomycin D, implying that de novo synthesis of tyrosinase was required.

While some correlation between the level of activity of tyrosinase in melanoma and clinical aggressiveness has been suggested (Pawelek, 1976), no conclusive evidence exists for such a relationship. MSH stimulation of the enzyme is associated with a decreased rate of cell proliferation in vitro with an interruption in the cell cycle at G_1 (Pawelek et al., 1975). This growth inhibition may be a direct consequence of the increased intracellular levels of cAMP associated with hormone treatment because the effects are mimicked by cAMP analogues, prostaglandin E_1 and cholera toxin (Pawelek, 1976).

Our laboratories have been involved in a study of the role of hormones and other small molecules as regulators of diverse functions in cultured cells. One topic of activity has been centered around the elucidation of low molecular weight factors which influence the expression of tyrosinase in cultured mouse melanoma cells. We have demonstrated that thyroid hormone and sex steroids can alter chemically-induced tyrosinase activity in concentrations which have physiological relevance.

METHODS AND RESULTS

B16/C3 mouse melanoma cultures were propagated in medium supplemented with fetal calf serum and in the absence of antibiotics. Initial studies established that none of the hormones added to rapidly replicating cultures affected growth curves. Tyrosinase activity was assayed according to the method of Pomerantz (1966).

Imidazole addition to culture medium can obviate the requirement for cAMP in adenyl-cyclase-deficient strains of E. coli (Kline et al., 1979) and the compound behaves in a manner analogous to the cyclic nucleotide in the induction of specific catabolite-repressible gene transcription (Kline et al., 1980). Because increased intracellular levels of cAMP elevate tyrosinase activity, imidazole was added to proliferating cultures of B16 cells. The compound induced the enzyme activity 4-5 times above control levels (Montefiori and Kline, 1981). The effect occurred within an hour of addition to the culture medium at concentrations in the millimolar range and was sustained for several hours. Imidazole exerted its stimulatory effect at a pre-translational level (Montefiori and Kline, 1981), on the basis of the studies utilizing inhibitors of RNA and protein synthesis. This induction was independent of alterations in intracellular cAMP content, unlike the effects of MSH.

Having defined a potent chemical inducer of the tyrosinase in B16 cells, the question of the naturally occurring hormones being significant regulators of induced and non-induced expression could be addressed. These seemed relevant because of a number of clinical and early experimental observations which suggest that the hormonal milieu might be a significant influence on melanogenesis and melanoma growth including: 1) Apparent sex differences in humans in regard to the clinical course of the disease (Chaudhuri et al., 1979b; Rampen and Mulder, 1980). 2) Anecdotal reports that thyroid hormone stimulated the growth and metastatic potential of tumors. 3) The finding of receptors for steroid hormones in a number of primary tumors and metastatic foci from individuals with early and advanced disease (Neifeld and Lippman, 1980).

Thyroid Hormone Regulation of Tyrosinase Activity

Initial studies dealt with an examination of the effect of the active thyroid hormone 3,3',5-L-triiodothyronine (T_3) on tyrosinase activity. Proliferating B16 cultures were treated with diluent, T_3 (10 nM), imidazole (10 mM) or both compounds for ∿ 18 hours and were then assayed for enzyme activity. As Figure 1 demonstrates, the hormone inhibited activity by ∿ 60%. The inactive hormone analogue 3,3',5'-L-triiodothyronine (RT_3) failed to influence activity. Imidazole could enhance the activity of the enzyme by ∿ 3-fold,

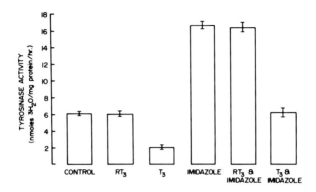

Figure 1. Effects of imidazole, T_3 and rT_3 on tyrosinase activity. Cultures were incubated with no additives (control), imidazole (10 mM), T_3 (10 nM), or rT_3 (10 nM) and allowed to proliferate for 19 h before harvesting and assaying for tyrosinase activity. Each bar represents the mean and range of duplicate cultures assayed for tyrosinase activity. (From Kline et al., 1986).

however T_3 markedly blocked this induction to levels which were indistinguishable from those in control cultures. In contrast, RT_3 did not alter imidazole induction at concentrations which were fully maximal for the T_3 effect. These hormone actions, like those of imidazole were quite rapid, occurring within 2 hours of addition to the culture medium (Fig. 2A). Within 8 hours of treatment, the T_3 effects began to disappear and activities in the hormone treated cultures rose. This increase was blocked by a re-addition and suggested that the hormone was disappearing from the medium. The most likely explanation was that the proliferating cultures were metabolizing T_3 at a rapid rate. The effects of T_3 were rapidly reversible, as Figure 2B demonstrates. Cultures were incubated in the presence of T_3 and imidazole for 6 hours. Eighteen hours later, half were shifted to medium supplemented with imidazole alone while the others were incubated in the continued presence of newly supplemented T_3 and imidazole. A detectable increase in enzyme activity occurred within 30 minutes of the shift in cultures treated with imidazole alone. Within 12 hours, these cultures attained activities approximating those of imidazole-treated cultures that had never received T_3 (Fig. 2A).

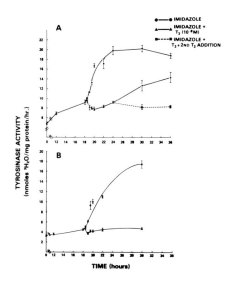

Figure 2. Time course of the effect of T_3 on imidazole-induced tyrosinase activity (A). Cultures were seeded in the presence of imidazole (10 mM) and were allowed to proliferate for 18 h. Some cultures were then shifted to medium containing both imidazole (10 mM) and T_3 (10 nM; ▲), while others were incubated with imidazole (●) alone. At 24 h, a set of the T_3-treated cultures were reinoculated with the hormone (■). In the experiments depicted in B, all cultures were incubated from initial plating with both imidazole (10 mM) and T_3 (10 nM). At 18 h, half of the cultures were shifted to medium containing only imidazole (●), while the others were again supplemented with both imidazole and T_3 (▲). Each datum point represents the mean ± range of duplicate cultures assayed for tyrosinase from one of three reproducible experiments (From Kline et al., 1986).

Maximal effects of T_3 were achieved at concentrations of 10 nM and half maximal responses occurred at ∿1 nM in experiments conducted in medium supplemented with 10% FCS (Fig. 3). Since the total T_3 concentration in FCS is 2.5 nM

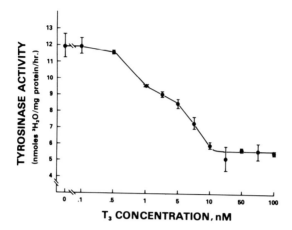

Figure 3. Effect of T_3 concentration on imidazole-induced tyrosinase activity. Freshly seeded cultures were treated with imidazole (10 mM) and various concentrations of T_3 (0-100 nM), allowed to proliferate for 19 h, then harvested and assayed for tyrosinase activity. Each datum point represents the mean ± range of duplicate cultures assayed for the enzyme from one of three reproducible experiments. (From Kline et al., 1986).

(Smith et al., 1982), the "free" hormone concentration under these culture conditions based upon equilibrium dialysis measurements should be 2 pM. Maximal inhibition of tyrosinase would appear to occur at a free concentration of 8 pM. Similar results were obtained in medium supplemented with thyroidectomized calf serum which contained < 20 ng/dl T_3 and < 0.3 ug/dl T_4. These results suggest that while the thyroid hormone effects involved the interaction of T_3 and a binding determinate which saturates at concentrations above those associated with euthyroidism, the hormone exerts a regulatory effect in the physiological range.

T_3 appears to require an intact cell for it to affect tyrosinase activity. When added to disrupted cell preparations which were then incubated for up to 4 hours, no influence of the hormone on enzyme activity could be demonstrated at concentrations which had maximally inhibited basal and imidazole-induced activity in culture. To determine the level of control at which T_3 was acting (i.e.

enzyme activation-inactivation, translation or pretranslation), a number of studies with inhibitors of protein and RNA synthesis were conducted. Cultures were pretreated with cycloheximide without or with T_3 for 5 hours, cell layers were washed extensively and covered with medium containing only actinomycin D. As Figure 4 demonstrates, the enzyme activity begins to increase after the shift to actinomycin D in cultures pretreated in the absence of hormone. Those cultures receiving T_3 failed to demonstrate an increase in activity during the entire duration of this study (5 hours). The rise in activity during actinomycin D treatment should reflect the translation of tyrosinase mRNA which had accumulated during cycloheximide pretreatment. The inference which we draw from the failure of this increase to occur

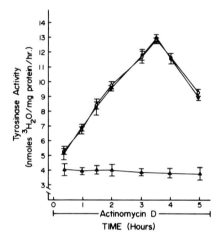

Figure 4. Protein and mRNA inhibitor effects on T_3 and imidazole regulation of tyrosinase activity. A set of 18 h imidazole-induced cultures was supplemented with cycloheximide (10 ug/ml; ●) for 5 h, a second set was treated with cycloheximide plus T_3 (10 nM; ▲), while a third set was incubated with cycloheximide plus rT_3 (10 nM; ○) and incubated for the same time period. After 5 h, medium was removed from each flask, and actinomycin D (1 ug/ml) was added. Quadruplicate cultures were harvested at the times indicated along the abscissa and assayed for tyrosinase activity. The error bars represent the mean ± range of activities from four independent assays for each datum point, and values represent one of two reproducible experiments. (From Kline et al., 1986).

in T_3 pretreated cultures is that the hormone may have decreased the accumulation of translatable mRNA. This, of course, could have been the consequence of either an alteration in the transcriptional rate or an effect on mRNA stabilization. To rule out the latter, a complementary experiment was performed in which T_3 was added, not with cycloheximide pretreatment but during the subsequent actinomycin D treatment. Cultures receiving T_3 exhibited an identical rise in enzyme activity to those receiving the antibiotic alone, suggesting strongly that the hormone does not alter the fate of preformed message and that the rate of tyrosinase mRNA transcription is being regulated.

In a recent set of experiments, it was evident that while T_3 can consistently and significantly inhibit tyrosinase expression, thyroxine (T_4) has no effect on melanogenesis in B16 cultures even at concentrations 100-fold higher than those necessary to demonstrate T_3 repression (Smith et al., 1987). This result implies that these cultures are not deiodinating T_4 to T_3 at a sufficiently rapid rate to produce adequate intracellular T_3 concentrations and offers, perhaps, direct evidence that T_4 may be without intrinsic biological activity in regard to some cellular metabolic events. The merits of these possibilities are currently under investigation in our laboratories.

The Role of Steroid Hormones in the Regulation of Tyrosinase

The results of studies designed to characterize the effects of T_3 on tyrosinase expression have provided the prototypic experiments for the assessment of steroid activity on melanogenesis in B16 culture. As Figure 5 suggests, the active estrogenic steroid 17β-estradiol (10 nM) had no effect on tyrosinase activity in proliferating cultures. When cultures were treated with imidazole, however, the estrogen could effectively block enzyme induction by that compound (Kline et al., 1987a) (Figure 6). The hormone did not affect the activity of preformed enzyme, consistent with the findings regarding T_3 repression. The inactive analogue, 17α-estradiol failed to block imidazole induction or to influence basal enzyme activity.

Preliminary results of inhibitor studies performed in a manner analogous to those examining thyroid hormone effects have yielded similar results: estradiol appears to attenuate

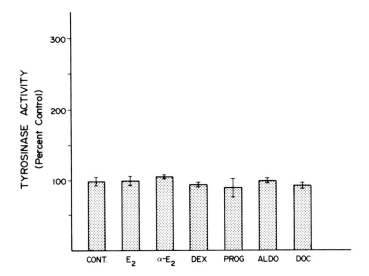

Figure 5. Effect of steroid hormones on tyrosinase activity. Cultures were incubated with the compounds indicated along the abscissa (10 nM) and were incubated for 19 h and analyzed for enzyme activity.

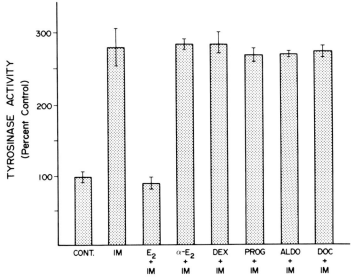

Figure 6. Effect of steroid hormones on imidazole-induction of tyrosinase. Cultures were treated as indicated in the legend to Figure 5.

the accumulation of translatable tyrosinase mRNA which is independent of any destabilization. Testosterone, but not 5α-dihydrotestosterone could also block imidazole induction at concentrations similar to those necessary for the estradiol blockade (Kline and Carland, 1987b).

While certain active estrogenic and androgenic steroid compounds could exert a significant repressive influence on tyrosinase induction by imidazole, Dexamethasone failed to alter enzyme activity. Similarly, progesterone, aldosterone and deoxycorticosterone (all 10 nM) were without effect on either basal (Figure 5) or imidazole-induced (Figure 6) tyrosinase activity, suggesting a degree of hormone specificity in regard to steroid class.

DISCUSSION

The findings to date from our laboratories suggest that T_3 but not T_4 can exert a significant inhibitory effect on basal tyrosinase activity in B16 melanoma cell cultures. The hormone can, in addition, block the induction of the enzyme by imidazole. These T_3-dependent effects are stereospecific, are dose-dependent and occur within a few hours of addition to culture medium. The results of studies involving inhibitors of RNA and protein synthesis suggest that T_3 blocks the accumulation of translatable tyrosinase mRNA through a mechanism which is independent of message destabilization, and that the action may be at the level of transcriptional control. Unlike T_3, 17β-estradiol and testosterone fail to influence basal enzyme activity but can block imidazole induction, apparently at a transcriptional level as well. Preliminary results utilizing a cDNA tyrosinase probe to quantitate hybridizable mRNA corroborate the results of the inhibitor studies.

The demonstration of high affinity binding sites for a hormone in a given mammalian tissue implies a potential importance for the hormone in some aspect of cellular metabolism. Thus the demonstration of steroid binding sites in these melanocyte tumors strengthened the presumption that sex and glucocorticoid steroids as well as other hormones participate in the regulation of melanoma growth. In fact, specific estrogen binding was found in 45% of the tumors examined in a series of 35 patients with melanoma, progesterone binding in 22%, and binding of glucocorticoids in 19%

(Fisher et al., 1976; Neifeld et al., 1976). This demonstration of high affinity, limited capacity binding does not however, necessarily imply that biologically functional receptors are present, as these authors are quick to point out (Neifeld and Lippman, 1980). The presence of specific estrogen binding has been confirmed by other investigators (Chaudhuri et al., 1979a). Several patients at the National Cancer Institute and elsewhere have been treated with either estrogens or anti-estrogens however, the responses are few and thus this therapeutic experience is quite disappointing (Neifeld and Lippman, 1980).

Studies conducted in cell culture thus far have yielded conflicting results. While one group failed to detect specific estrogen binding in any of 21 human melanoma cell lines tested (Neifeld and Lippman, 1980), another found estrogen and progesterone binding activity in 4 of 6 and 3 of 6 lines, respectively as well as a hormone response in those lines with binding activity (Chaudhuri et al., 1979b).

Glucocorticoid receptors are present in a majority of human melanomas (Bhakoo et al., 1981a), in an established human melanoma cell line (Di Sorbo et al., 1983) and in a variety of murine culture lines including B16 (Abramowitz and Chavin, 1978; Bhakoo et al., 1981b). In addition to this evidence for receptors, other investigators have demonstrated effects in these models including growth inhibition (Di Sorbo et al., 1983; Bhakoo et al., 1981b) and inhibition of tyrosinase activity (Abramowitz and Chavin, 1978). We were unable to demonstrate an effect of this class of steroid hormone in B16 cultures. The basis for the apparent discrepancies between these findings and our own is not readily apparent but might reflect differences in culture conditions, endogenous levels of hormone in serum, or changes in the B16 cell line which may have occurred spontaneously.

Melanogenesis in B16/C3 cultures is thus apparently under multi-hormonal control, consistent with a variety of other cellular metabolic events already reported in other tissues, including α_{2u} globulin gene expression (Kurtz and Feigelson, 1978; Roy et al., 1976), growth hormone synthesis (Martial et al., 1977) and hyaluronate synthesis in human skin fibroblast cultures (Smith, 1984). Whether the action of one hormone on tyrosinase activity can influence that of another (such as synergism) is not yet known. This model may be ideal for studying the nature of hormone-hormone interactions at the level of gene expression and for examining the mechanisms involved in hormonal conditioning of the cellular response to chemicals, as has been described previously in liver (Sassa and Kappas, 1978; Smith et al., 1982). In addition, inferences regarding the complex interactions of small molecules at the level of DNA at specific control regions, metabolite gene regulation, may be made from further studies which parallel those conducted in procaryotic cells (Kline et al., 1979). Whether these regulatory effects are mediated through receptors is as yet not known.

Experimentation performed in vitro offers important advantages over that conducted in either whole animals or humans. It offers the opportunity to assess the biological role of regulators of metabolism in isolated, homogeneous cell populations under stringently controlled experimental conditions. Because the behavior of cells maintained in culture may differ from that in situ in fundamental ways, observations made in these models should complement rather than replace those conducted in whole animals. Thus, the results reported here ultimately need to be extended by further studies in isolated tissue preparations and in the intact organism.

ACKNOWLEDGEMENTS

The authors with to thank K. Carland for her participation in these studies and R. Harvey for preparing the manuscript. This work was supported in part by the Veterans Administration Research Service and ELK-PAM Research Fund.

REFERENCES

Abramowitz J, Chavin W (1978). In vitro effects of hormonal stimuli upon tyrosinase and peroxidase activities in murine melanomas. Biochim Biophys Res Commun 85:1067-1073.
Abramowitz J, Chavin W (1978). Glucocorticoid modulation of adrenocorticotropin-induced melanogenesis in the Cloudman S-91 melanoma in vitro. Exp Cell Biol 46:268-276.
Bhakoo HS, Milholland RJ, Lopez R, Karakousis C, Rosen F (1981a). High incidence and characterization of glucocorticoid receptors in human malignant melanoma. J Natl Canc Inst 66:21-25.

Bhakoo HS, Paolini NS, Milholland RJ, Lopez RE, Rosen F (1981b). Glucocorticoid receptors and the effect of glucocorticoids on the growth of B16 melanoma. Canc Res 41:1695-1701.
Chaudhuri PK, Walker MJ, Keehn D, Beattie CW, Das Gupta TK (1979a). Estrogen receptor in human benign nevi and skin. Surgical Forum 30:128-129.
Chaudhuri PK, Walker MJ, Beattie CW, Das Gupta TK (1979b). Endocrine correlates of human malignant melanoma. J Surg Res 26:214-219.
Chaudhuri PK, Das Gupta T, Beattie CW, Walker MJ (1982). Glucocorticoid-induced exacerbation of metastatic human melanoma. J Surg Onc 20:49-52.
Coppock DL, Straus DS (1983). Complementation of the growth response to insulin and MSA occurs at a step distal to hormone-receptor interaction in mouse embryo fibroblast x melanoma cell hybrids. J Cell Physiol 114:123-131.
Danforth DN, Russell N, McBride CM (1982). Hormonal status of patients with primary malignant melanoma. So Med J 75:661-664.
Di Sorbo DM, McNulty B, Nathanson L (1983). In vitro growth inhibition of human malignant melanoma cells by glucocorticoids. Cancer Res 43:2664-2667.
Fisher RI, Neifeld JP, Lippman ME (1976). Oestrogen receptors in human malignant melanoma. Lancet 1976 ii:337-340.
Fuller BB, Viskochil DH (1979). The role of RNA and protein synthesis in mediating the action of MSH on mouse melanoma cells. Life Sciences 24:2405-2415.
Kline EL, Bankaitis V, Brown CS, Montefiori D (1979). Imidazole acetic acid as a substitute for cAMP. Biochim Biophys Res Comm 87:566-574.
Kline EL, Brown CS, Bankaitis V, Montefiori DC, Craig K (1980). Metabolite gene regulation of the L-arabinose operon in Escherichia coli with indolacetic acid and other indole derivatives. Proc Natl Acad Sci (USA) 77:1768-1772.
Kline EL, Carland K, Smith TJ (1986). Triiodothyronine repression of imidazole-induced tyrosinase expression in B16 melanoma cells. Endocrinol 119:2118-2123.
Kline EL, Carland K, Blackmon B (1987a). Pre-translational inhibition of tyrosinase by estradiol and estriol in B16 melanoma cells. (Submitted for publication).
Kline EL, Carland K (1987b). Androgenic control of imidazole induction of tyrosinase in B16 melanoma cell culture. (Submitted for publication).

Kurtz DT, Feigelson P (1978). Multi-hormonal control of messenger-RNA for hepatic protein alpha-2u globulin. Biochem Actions Horm 125:433-455.

Lee TH, Lee MS, Lu M-Y (1972). Effects of MSH on melanogenesis and tyrosinase of B16 melanoma. Endocrinol 91:1180-1188.

Martial JA, Seeburg PH, Guenzi D, Goodman HM, Baxter JD (1977). Regulation of growth hormone gene expression. Synergistic effects of thyroid and glucocorticoid hormone. Proc Natl Acad Sci (USA) 74:4293-4295.

Montefiori DC, Kline EL (1981). Regulation of cell division and of tyrosinase in B16 melanoma cells by imidazole: a possible role for the concept of metabolite gene regulation in mammalian cells. J Cell Physiol 106:283-291.

Neifeld JP, Lippman ME (1980). Steroid hormone receptors and melanoma. J Invest Derm 74:379-381.

Pawelek J, Wong G, Sansone M, Morowitz J (1973). Molecular controls in mammalian pigmentation. Yale J Biol Med 46:430-443.

Pawelek J, Sansone M, Koch N, Christie G, Halaban R, Hendee J, Lerner AB, Varga JM (1975). Melanoma cells resistant to inhibition of growth by stimulating hormone. Proc Natl Acad Sci (USA) 72:951-955.

Pawelek J (1976). Factors regulating growth and pigmentation of melanoma cells. J Invest Derm 66:201-209.

Pomerantz SH (1966). The tyrosine hydroxylase activity in mammalian tyrosinase. J Biol Chem 241:161-168.

Rampen FHJ, Mulder JH (1980). Malignant melanoma: An androgen-dependent tumour? Lancet 1980:562-565.

Roy AK, Schiop MJ, Dowbenko DJ (1976). The role of thyroxine in the regulation of translatable messenger RNA for α_{2u} globulin in rat liver. FEBS Lett 64:396-399.

Sassa S, Kappas A (1978). Induction of δ-aminolevulinate synthetase and porphyrins in cultured tumor cells maintained in chemically defined media. J Biol Chem 252:2428-2436.

Smith TJ, Murata Y, Horwitz AL, Philipson L, Refetoff S (1982). Regulation of glycosaminoglycan synthesis by thyroid hormone in vitro. J Clin Invest 70:1066-1073.

Smith TJ, Drummond G, Kourides IA, Kappas A (1982). Thyroid hormone regulation of heme oxidation in the liver. Proc Natl Acad Sci (USA) 79:7537-7541.

Smith TJ (1984). Dexamethasone regulation of glycosaminoglycan synthesis in cultured human skin fibroblasts: Similar effects of glucocorticoid and thyroid hormones. J Clin Invest 74:2157-2163.

Smith TJ, Warren J, Kline EL (1987). Triiodothyronine but not 3,3',5,5'-tetraiodothyronine regulates tyrosinase expression in murine melanoma in vitro. (In preparation).

Wong G, Pawelek J (1973). Control of phenotypic expression of cultured melanoma cells by melanocyte stimulating hormones. Nature (New Biology) 241:213-215.

Wong G, Pawelek J (1975). MSH promotes the activation of preexisting tyrosinase molecules in Cloudman S91 melanoma cells. Nature (Lond) 255:644-646.

HORMONE ASSOCIATED THERAPY OF LEUKEMIA: REFLECTIONS

W. David Hankins, Kyung Chin and George Sigounas

Armed Forces Radiobiology Research Institute/
Laboratory of Chemical Biology
National Institutes of Health
Bethesda, Maryland 20205

Human leukemias are generally considered to be clonal disorders in which entire leukemic cohorts arise from single transformed cells. They are often characterized according to a specific hemopoietic cell lineage (granulocytic, erythrocytic and lymphocytic types) or whether the disease manifests as a chronic or an acute disease. The majority of leukemias are thought to arise from committed, unipotent blood cell progenitors since the leukemia cells are all within a single lineage. However, others such as chronic myelogenous leukemia (CML) appear to emanate from a transformed multipotent stem cell as the leukemic cells exhibit three or four hemopoietic phenotypes.

Current therapies for leukemia include radiation, chemotherapy, bone marrow transplantation or combinations of these modalities. In general, these treatments are not very effective and the majority of patients who develop leukemia will die from the disease.

In an effort to force leukemia cells to die through maturation, several laboratories are currently attempting to find chemicals or hormones which will induce the "maturation arrested" cells to differentiate. It should be emphasized, however, that there is disagreement amongst oncologists as to whether leukemic cells are actually blocked in maturation. Conceptionally, this is important because if maturation is not blocked, the leukemic cohort could by physiologic processes, undergo clonal extinction by terminal differentiation if given enough time. Another unsettled issue with therapeutic implications is whether cancer cells retain their requirements for physiologic hormones. If leukemic cells do remain hormone-dependent, they might be eradicated by deprivation of essential hormones. It is obvious then, that the answers to these questions could provide new direction for the development of more effective therapeutic tools with which to combat leukemia.

Over the past several years, we have studied retrovirus-induced leukemia in mice. Taken collectively, our data strongly suggest that animal leukemias retain hormone dependence and in many cases, the transformed cells are not blocked in maturation. Of greater significance, we have recently observed that certain leukemias responded dramatically to hormone related therapy.

Our interest in the retrovirus induced leukemias stemmed from our desire to further understand the developmental regulation of red blood cell formation. Erythropoietin (epo) is a glycoprotein hormone, which is generally accepted as the primary physiologic regulator of erythropoiesis. In 1971, Dr. Charlotte Friend isolated some erythroleukemia cell lines which appeared to grow independently of erythropoietin. The leukemia cells appeared to be arrested in maturation. However, when exposed to a variety of non-physiologic agents, such as dimethylsulfoxide and butyric acid, the cells underwent terminal erythroid differentiation.

Our initial goal was to gain insight into the molecular events which mediated the transition of normal erythroid cells into leukemic cells which apparently did not need the hormone for proliferation and which seemed to be blocked in maturation. We were fortunate to develop an in vitro transformation system in which direct exposure of bone marrow to certain retroviruses led to the appearance of "transformed" erythroid cells in culture (1). The acquired growth advantage of the transformed cells permitted them to form colonies in methylcellulose cultures whereas normal erythroid progenitors would not grow, i.e. without added erythropoietin. Subsequent studies employing this in vitro transformation system led us to the surprising observations that the virus treated cells were neither blocked in maturation nor hormone-independent. Indeed, rather than being epo-independent, these cells were actually hypersensitive to the hormone. Thus, these virus transformed cells appeared to require only the minute amounts of epo which were present in the fetal calf serum included in the culture medium. In subsequent studies, Koury and Bondurant (2) have exploited this epo "hypersensitivity" of the virus treated cells as a useful model in which to study the terminal events of hormone-mediated erythroid differentiation.

Initially, it was our feeling that the hormone sensitivity of the erythroleukemia cells induced by Friend virus might be unique to this particular retrovirus. However, in subsequent studies we observed that a large number of oncogenic retroviruses transformed erythroid, lymphoid and myeloid cells which retained hormone sensitivity and continued to undergo developmental maturation (3). Thus, the general picture which emerged was that transformed cells could gain a heritable growth advantage without losing their ability to differentiate and without losing their dependence on physiologic hormones.

Since these observations were made in vitro, we decided to test whether erythroleukemias might also remain hormone-dependent in vivo. As an experimental model, we chose an aggressive transplantable erythroleukemia induced by a replication-competent retrovirus isolated and cloned by Oliff et al (4). Upon transplantation of leukemic spleen cells into normal adult mice, the recipients develop fatal erythroleukemia within six to eight weeks. When we tested the leukemic cells for hormone sensitivity in culture, we found them to be absolutely dependent on erythropoietin. Figure 1 shows a typical experiment in which the spleen cells form the leukemic mice were cultured in methylcellulose (5). Without erythropoietin, no colonies were detected. With erythropoietin, more than 100 colonies per plate were observed. The hormone dependence was lineage specific since other hemopoietic colony stimulated factors would not substitute for erythropoietin. The apparent absolute dependence of colony formation on the hormone of leukemic cells was surprising, and suggested two possibilities which are currently under test in our laboratory. First, the

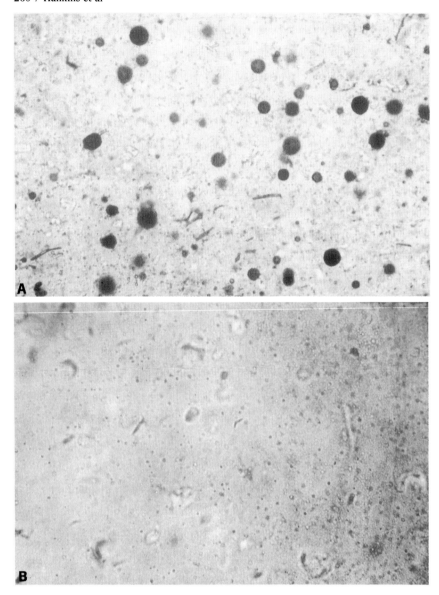

Fig. 1. Effect of EPO on leukemia cell growth in vitro. Spleen cells from passage 5 of transplantable erythroleukemia line 19 were seeded at a concentration of 1 million cells/ml in methylcellulose as described in Materials and Methods. Connaught step III erythropoietin was added to half the cultures (A) and the remainder (B) received an equivalent amount of Hanks' balanced salt solution. Photographs were made in situ using a Leitz inverted microscope at a magnification of X25.

dependence of the erythroleukemia cells on erythropoietin for viability suggests that the hormone may serve as a permissive factor to keep committed erythroid progenitors alive rather than acting to induce uncommitted stem cells to traverse the erythroid pathway. Second, we are testing whether the erythropoietin dependence may be exploited in vivo to develop an anti-hormone therapy for the leukemia. Although further testing is essential, our preliminary results are very encouraging and are presented below.

First, with respect to the possibility that erythropoietin may act as a viability factor, we have sub-cultivated colonies similar to those in figure 1A. Colonies were dispersed in culture media (without methylcellulose) containing IMDM and 30% fetal calf serum. In the experiment shown in table 1, the cultures were seeded with and without erythropoietin every three days for nine passages. At each passage, the cells growing with erythropoietin were washed and re-suspended at 1/4 the cell concentration. The results indicated that erythropoietin was essential throughout the experiment and without the hormone these erythroleukemia cells quickly died. We are currently attempting to establish long term EPO-dependent cell lines which should be of value for receptor and mechanistic studies. In any case, while these results to not rule out the possibility that erythropoietin may still induce erythroid differentiation, they do strongly emphasize the hormone's role in maintaining viability and cell proliferation.

As to the possibility that the leukemic cells might be controlled in vivo by endocrine therapy, we sought ways to lower endogenous erythropoietin. Since only limited quantities of neutralizing erythropoietin antiserum were available, we employed red blood cell transfusion as a physiologic feedback means of lowering the hormone.

To test the effect of epo deprivation on leukemia growth, 50,000 leukemic cells were inoculated into mice and the recipients were hypertransfused at weekly intervals. As can be seen from the comparison of groups A and B in figure 2, the hypertransfusion led to a significant extension (P=0.001) in the survival times of the leukemic mice. Thus, the mice that were transfused lived two to three times longer than those that did not

TABLE 1

Dependence of subcultured erythroleukemia colonies on EPO

CELL/ml × 10^6

Passage Number	1	2	3	4	5	6	7	8	9
EPO Present	2.00	1.10	1.50	0.85	1.95	1.00	2.00	0.70	1.75
EPO Absent	0.05	0.05	0.05	0.10	0.25	0.70	0.30	0.10	0.10

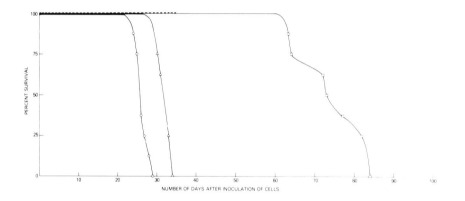

Fig 2. Effect of hypertransfusion or EPO administration on the survival of mice with erythroleukemia. One million cells from transplantable erythroleukemia line EL-19 were injected intravenously into each 50 adult mice. The mice were divided into four groups. Group A (△) received no further treatment. Group B (□) received weekly transfusions of 1 ml per mouse of reconstituted blood containing 80% red blood cells. Group C (○) was transfused as B, but each mouse also received daily intraperitoneal injections of 0.2 unit of Connaught step III erythropoietin. All mice were monitored and the spleens palpated each day. Group D (---) was a control for group C. These mice received daily injections of EPO but had not been inoculated with leukemic cells. None of the mice in group D died during the monitoring period.

receive red blood cells.

The increased survival times were most likely mediated through a reduction of endogenous epo brought about by the transfusions. However, since the leukemic mice were severely anemic, we considered the possibility that hypertransfusion had simply corrected the anemia and thereby increased survival. Two experimental findings argued against this contention. First, a reduced rate of growth of the leukemic cells was indicated by our finding that the spleens from transfused mice were considerably smaller than those of untreated mice. This difference in spleen size can be seen in figure 3. Second, simultaneous administration of erythropoietin plus hypertransfusion led to a more rapid leukemia and survival times similar to untreated recipients (group B and C, for example, were significantly different at P=0.001). Thus while the transfusions corrected the anemia, these treatments were of little survival value when mice simultaneously received the hormone. We believe it is likely therefore that the retardation of leukemia growth by hypertransfusion is due to a deprivation of erythropoietin.

Fig. 3. Effect of transfusion of spleen size in erythroleukemic mice. On day 25 of the experiment described in the legend to Figure 2, one mouse from group A (bottom, hypertransfused) and one from group B (top, untreated) were sacrificed, and the spleens and livers were exposed and photographed.

These encouraging results in the late phase of the disease prompted us to reassess the effectiveness of hypertransfusion at an earlier time, - i.e. during leukemia development. Newborn mice were inoculated with virus and erythroid hyperplasia developed in 100% of the mice. At five weeks after virus inoculation, the animals exhibited severe anemia and greatly enlarged spleens. At this time, the mice were divided into two groups. One group received weekly red blood cell transfusions (as in Figure 2) for eight consecutive weeks and the other group received mock intraperitoneal injection. As figure 4 shows, the effect of hypertransfusion was dramatic. Over a six month period, 70% of the treated mice survived while all of the controls were dead within weeks. It is of considerable interest that the treated mice survived long after hypertransfusion was discontinued. We believe this suggests the possibility that, as we saw in vitro,

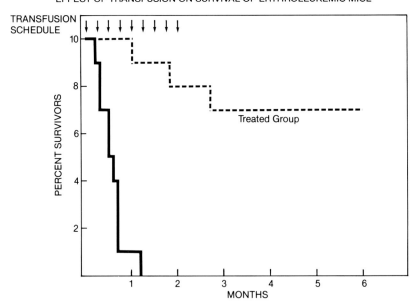

Fig. 4. Effect of transfusion on the survival of mice during the dearly stages of leukemia development. Treated groups received transfusions at times indicated by the arrows. Transfusions were administered as described in Figure 2.

the developing leukemia cells require erythropoietin for their viability and in its absence, do not survive. Undoubtedly, the normal erythropoiesis is also dependent on erythropoietin and is also suppressed by the transfusion. However, it is likely that when the transfusions were discontinued, new erythroid progenitors arose from more primitive stem cells and therein supported normal erythropoiesis. The leukemic clones, on the other hand were assumed to be either extinct or severely reduced in number and did not recover during the experiment.

A LOOK AHEAD

The experiments referred to and reported herein encourage further pursuit of an endocrine approach to the therapy of leukemia in experimental models. In addition, these in vivo therapeutic studies underscore the importance of current in vitro studies to define hormone sensitivity in a variety of human leukemias. For example, several labs (6-10) have recently documented the sensitivity of myeloid leukemia cells to human colony stimulating factors. Finally, the articles of Saharabuddle and Waldman in this volume strongly implicate the influence of hormones on lymphocytic leukemias.

Thus, some general questions may be timely and relevant in consideration of new approaches to treatment of human leukemias. For example, are all normal hemopoietic pathways dependent on specific viability factors? Are leukemic cells within these pathways also dependent on such regulators? Are the differentiated cells observed in chronic myelogenous leukemia reflections of a transformed multi-potential cell committed to terminal differentiation? If so, it is possible that appropriate hormonal deprivation might permit extended patient survival and ultimately allow extinction of the leukemic clone? Would inhibition of GM-CSF retard or halt the growth of acute myelogenous cells?

If the answers to such questions further implicate hormones in tumor growth, there will be an increased need for improved anti-hormone technology. While transfusion of red blood cells was an effective means for lowering erythropoietin in our experimental model, this method obviously has limited value in human leukemias because of logistical and histocompatibility problems. We and other labs have therefore begun to explore several alternative methods of suppressing hormones or interfering with their actions. Indeed, we have coined the operational term of "Hormone Associated Therapy" or HAT to encompass a number of therapeutic approaches that, we hope, will ultimately exploit the hormone sensitivity of tumor cells for the eradication or suppression of various cancers. The HAT approach includes the use of new culture methods and recombinant DNA technology to: (A) produce mutated hormone antagonists, (B) generate monoclonal antibodies to hormones or their receptors, (C) link toxic molecules or radionuclides to hormones as "guided missiles", or to (D) use viability hormones to selectively stimulate tumor cells to selectively increase their susceptibility to chemotherapeutic agents or radiation.

In summary, as each month brings about advances in molecular technology, new observations of hormone dependency in various leukemias and better understanding of endocrine regulation, we are optimistic that the HAT approach may, in the near future, remove leukemia as a major cause of death.

References:

1. W.D. Hankins, S.B. Krantz, T.A. Kost and M.J. Koury- In Vitro Transformation of Mouse Hemopoietic Cells by the Spleen Focus-Forming Virus. Cold Spring Harbor Symposia on Quantum Biology, Vol. 44, 1979.

2. M.J. Koury, M.C. Bondurant, D.T. Duncan, S.B. Krantz and W.D. Hankins - Specific Differentiation Events Induced by Erythropoietin in Cells Infected In Vitro with the Anemia Strain of Friend Virus. Proc. Natl. Acad. Sci. USA 79:635-639, 1982.

3. W.D. Hankins and J. Kaminchik - Modification of Erythropoiesis and Hormone Sensitivity by RNA Tumor Viruses. In: N. Young, R.K. Humphries, and A. Levine (Eds.): Aplastic Anemia, Stem Cell Biology and Recent Therapeutic Advances. New York, Alan R. Liss, Inc., 1984, pp. 141-152.

4. A. Oliff, S. Ruscetti, E.C. Douglas, E.M. Scolnick-

Isolation of Transplantable Erythroleukemia Cells from Mice Infected with Helper-Independent Friend Murine Leukemia Virus. Blood 58:244-254, 1981.

5. A. Hossain, J.K. Kim and W.D. Hankins. Treatment of a fatal transplantable Erythroleukemia by Procedures which Lower Endogenous Erythropoietin. J. Cellular Biochemistry 30:311-318, 1986.

6. D.C. Young and J.D. Griffin - Autocrine Secretion of GM-CSF in Acute Myeloblastic Leukemia. Blood 68:1178-1181, 1986.

7. M.C. Tweeddale, B. Lim, N. Jamal, J. Robinson, J. Zalchery, G. Lockwood, M. Minden and H.A. Messner - The Presence of Clonogenic Cells in High-Grade Malignant Lymphoma: A Prognostic Factor. Blood 69:1307-1314, 1987.

8. C. Kelleher, J. Miyanchi, G. Wong, S. Clark, M. Minden, and E.A. McCulloch - Synergism Between Recombinant Growth Factors, GM-CSF and G-CSF, Acting on the Blast Cells of Acute Myeloblastic Leukemia. Blood 69:1498-1503, 1987.

9. D. Metcalf, Review: The Molecular Biology and Functions of the Granulocyte-Macrophage Colony-Stimulating Factors. Blood 67:257-267, 1986.

10. M. Tomonaga, D.W. Golde and J.C. Gasson - Biosynthetic (Recombinant) Human Granulocyte-Macrophage Colony-Stimulating Factor: Effect on Normal Bone Marrow and Leukemia Cell Lines. Blood 67:31-36, 1986.

HUMAN B CELL GROWTH FACTOR AND NEOPLASIA

Chintaman G. Sahasrabuddhe, Sudhir Sekhsaria,
Barbara Martin, Linda Yoshimura and R. J. Ford
UNIVERSITY OF TEXAS SYSTEM CANCER CENTER
M.D. ANDERSON HOSPITAL & TUMOR INSTITUTE
1515 HOLCOMBE
HOUSTON, TEXAS 77030

INTRODUCTION

Three successive processes, activation, proliferation and differentiation are believed to lead a circulating B lymphocyte to mature antibody producing plasma cell. There is ample evidence to support the involvement of antigen independent soluble factors in modulation of some of these processes. A soluble factor known as B cell growth factor (BCGF), plays a crucial role in the proliferation of activated B cells [Muraguchi and Fauci 82, Maizel et al. 82, Yoshizaki et al. 83, Maizel et al. 83]. Several laboratories have reported multiple forms of human BCGF [Ambrus Jr. et al. 85, Shimizu et al. 85, Jurgensen et al. 86, Dugas et al. 85, Sahasrabuddhe et al. 84, 86]. Two major forms of human BCGF, a low M_r (12K-20K) and a high M_r (50K-60K) have been characterized to some extent. It is still unclear if these two forms are related to each other.

Our own studies have shown that the low M_r form is secreted from lectin activated T lymphocytes [Maizel et al. 82a]. While the high M_r form is localized within the intracellular compartment [Sahasrabuddhe 84]. Recently we have shown that neoplastic B cells from hairy cell leukemia (HCL) patients contain a cytosolic high M_r protein which exhibit BCGF activity on autochthonous HCL cells as well as on normal BCGF dependent human B cell lines [Ford et al. 86]. A variety of lymphoblastoid cell lines of neoplastic origin have been shown to secrete high M_r BCGF [Gordon et al., 84, Blazar et al. 83, Ambrus Jr. et al. 85, Shimizu et al. 85, Sahasrabuddhe et al. 87, Okada et al. 83]. In this report we present partial characterization of high M_r human BCGF and demonstrate its efficacy on BCGF dependent "normal" human B cell line as well as on the neoplastic B cells. We also discuss the possible involvement of high M_r form of BCGF (regarding its secretion and efficacy) as an autocrine or autochthonous stimulator of neoplastic B lymphocytes.

MATERIALS AND METHODS

Cells and Cell Lines: Fresh human B lymphocytes are isolated by negative selection from peripheral blood lymphocytes (PBLs) (Ford et al. 81). B cell growth factor dependent B cell lines derived from normal B cells are established as described earlier (Maizel et al. 83). B cell lines derived from B lymphoma patients are maintained in cultures in our laboratory (Ford et al. 85).

Preparation of cytosolic extracts: Human PBLs isolated from multiple donors were pooled and cultured at $5x10^6$ per ml of RPMI 1640 medium supplemented with 2.0mM glutamine, 0.25% (W/V) bovine serum albumin (BSA), 0.5% (V/V) phytohemaglutinin-M (PHA-M, GIBCO Laboratories, Grand Island N.Y. 14072), 100U/ml of penicillin and 100 µg/ml of streptomycin 4 (Maizel et al. 82). At the end of 16 hr. culture period cells were recovered by centrifugation, the cell pellet was suspended in hypotonic buffer and lysed by homogenization as described earlier (Sahasrabuddhe 84). The lysate was cleared of nuclei and mitochondria by centrifugation and the clarified lysate was centrifuged at 100,000xg in a 60 Ti rotor for 90 minutes to remove cellular membrane fragments and polysomal material. The clear supernatant was used as a source of cytosolic proteins for the purification of 60K BCGF. Similar cytosolic extracts were also prepared form the nonactivated freshly prepared human PBLs and from the neoplastic B cells.

Purification of 60k BCGF: 60K BCGF from normal human PBLs was purified to homogeneity by multi-step biochemical procedure including ammonium sulfate precipitation, isocratic gel filtration, ion-exchange chromatography and isoelectric focusing. [Sahasrabuddhe et al 86]. Homogeneity of isoelectrically focused material was tested by electrophoresis under denaturing conditions on sodium dodecyl sulfate-polyacrylamide gel [Laemmli 70] followed

by silver staining [Okley et al 80]. Tumor cell
supernatants and cytosolic extracts from HCL cells were
chromatographed on a sephacryl 200 isocratic gel
filtration column (2x100 cm) that had been calibrated with
variety of mol. wt. protein markers. The column elution
buffer was phosphate buffered saline (PBS, pH 7.0) and
fraction volume was 5 ml. The proliferative activities of
eluted fractions were tested on growth factor dependent B
cell lines.

Biological assays: The biological assay for BCGF activity
was routinely performed in 96-well flat bottom
microculture plates. Human B lymphocytes from growth
factor dependent long term B cell lines [Maizel et al 83]
were cultured in a final volume of 200 µl in RPMI 1640
medium supplemented with 2% heat inactivated fetal calf
serum or 1% Nutridoma-Hu (Boehringer Mannheim
Biochemicals, Indianapolis, IN 46250). A total of $1x10^4$ -
$1.5x10^4$ cells were used per well. The cultures were
incubated for 40 hrs in a humidified incubator purged with
5% CO_2. Sixteen hr. prior to harvest, the cells were
labelled with 1 µCi [^{3}H]-Tdr (sp. activity 6 Ci/mmoles),
and the incorporated radioactivity was measured by
scintillation counting.

RESULTS

Purification of intracellular 60K protein from normal
human PBLs: Table 1 summarizes the results of various
purification steps. The homogeneously pure protein, with
an overall purity of ≥99%, represented 0.075% of the
initial protein mass in the crude cytosolic extract.
Units of BCGF activity in samples at various stages of
purification were calculated by probit analysis [Gillis et
al.]. A partially purified preparation of low M_r BCGF
(Cellular Products, Inc. Buffalo, N.Y. 14202), devoid of
any IL-2 or IL-1 activities, was used as a standard in
microculture assays utilizing long-term B cell lines.
This comparison did not necessarily reflect absolute
values of intracellular BCGF in a sample for the reason
that the inducive capacity of the intracellular BCGF may
not be same as that of the low molecular weight BCGF. Yet
this comparison provided a basis for calculating relative
yields of intracellular BCGF. Thirty three percent of the

PURIFICATION TABLE 1

Purification Step	Total optical density units at 230 nm	Protein[a] Recovery (%)	Total BCGF[b] Units	Yield (%)
Crude Extract	135.0	100	14354	100
Ammonium Sulfate Precipitation	53.4	39.6	8955	62.4
Sephacryl-200 isocratic gel filtration	15.75	11.7	5833	40.6
DEAE-ion exchange	4.5	3.3	5000	34.8
Isoelectric focusing	0.1	.07	4762	33.1

a. The protein concentration in a sample was monitored by optical density (OD) at 230nm, and the percent recovery at each purification step was calculated by comparing total OD units with initial OD units.

b. A unit of BCGF activity is defined by its capacity to induce 50% of maximal proliferative response in B cells in microculture assays.

BCGF activity in the starting material was recovered as a homogeneously pure protein, with an estimated specific activity of 10^5 U/mg.

Figure 1 shows the results of isoelectric focusing at the final step of purification. The pool of active material from the previous step of purification (Table 1; DEAE-sepharose ion exchange chromatography) was isoelectrically focused on a preparative sucrose gradient in the presence of ampholytes (LKB) covering a pH range of 3.5-9.5. The biological activity was found reproducibly in the pH range of 6.1-6.5 pH units. This isoelectrically focused material was analyzed by electrophoresis on sodium-dodecil sulfate-polyacrylamide gel under denaturing conditions. Silver stained gel indicated presence of a single protein band at 60K (figure 2).

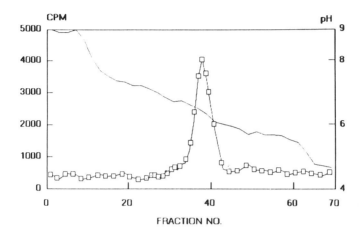

Figure 1. Isolectric focusing of the biologically active 60k protein. Pooled fractions from DEAE-ion exchange chromatography were dialyzed against 10% glycerol and isoelectrically focused on a sucrose density gradient containing 1.2% (v/v) ampholytes covering 3.5-9.5 pH range. Electrophoretic focusing was conducted at constant power of 15 watts, and the process was terminated when the initial current of 16 mA dropped to the level of 1-2 mA. The gradient was fractionated by collecting 2 ml per fraction. Every alternate fraction was tested for BCGF activity in microculture assay utilizing BCGF dependent B cells. Continuous line shows the pH values and the open squares show the BCGF activity monitored by ^{3}H-Tdr incorporation.

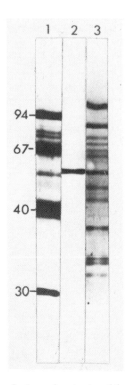

Figure 2: Analysis of isoelectrically focussed 60k protein by SDS-polyacrylamide gel electrophoresis and silver staining. Ten nanograms of isoelectrically focused biologically active 60k protein and 1 µg of prefocused protein sample were analyzed by SDS-PAGE on a 10% acrylamide gel, followed by sensitive silver staining procedure.

Lane 1: Molecular weight markers; molecular weights in thousands of daltons are designated.

Lane 2: Isoelectrically focused 60k protein.

Lane 3: Preisoelectrical focusing proteins sample containing 60k protein.

Biological activity of homogeneously purified 60K protein: It has been reported that the high M_r BCGF synergizes with dextran sulfate and not with anti-IgM, in a proliferative assay using freshly isolated human B lymphocytes [Okada et al 83]. Our preliminary data suggested that the crude preparations of cytosolic 60K protein exhibited barely detectable levels of BCGF activity in anti-IgM costimulatory assays. Therefore, we tested the biological efficacy of homogeneously purified intracellular 60K protein in both anti-IgM and dextran sulfate costimulatory assays utilizing freshly isolated human B cells. The low M_r BCGF demonstrated significant biological activity with both activators. However, no BCGF activity was observed when 60K protein was used with either activator [Table 2]. In contrast long term B cells responded equally well to both the low as well as high M_r BCGF.

Production of Growth Factor by HCL cells: Cells from patients showing leukemic form of HCL disease ($\geq$20,000 WBC/µl; $\geq$80% leukemic cells by morphology) were collected either by leukapheresis or by venipuncture. Phenotypic characterization of HCL cells indicated standard B cell membrane markers along with some myeloid-monocytic and a Tac antigen [Ford et al. 86]. These HCL cells were cultured in vitro in either low serum (1%) or serum-free (0.25% BSA) conditions for 24 hrs in a humidified incubator at 37°C purged with 5% CO_2. The supernatants harvested at the end of culture period were assayed for BCGF activity using either long-term growth factor-dependent B cells or autochthonous HCL cells. Results of a typical proliferative assay are shown in Table 3. These results demonstrate that spontaneous release of BCGF molecules is associated with neoplastic B lymphocytes.

In our earlier studies (Sahasrabuddhe et al. 84), we have shown that 60k protein is not present in the cytoplasm derived from normal human B lymphocytes. Therefore, we assayed the cytoplasmic extracts of the B cells from hairy cell leukemia patients. As shown in table 4 these extracts contained low but significant amount of BCGF activity.

TABLE 2

BIOLOGICAL ACTIVITY OF LOW M_r BCGF AND 60K PROTEIN

Additions	Concentration (% v/v)	Human G_0B cells* Anti-Igm S.I.*	Dextran Sulfate S.I.	B cells from BCGF dependent B cell line S.I.
Low M_r BCGF	1.2	---	---	21.5
	2.5	9.2	8.8	26.4
	5.0	11.1	11.03	33.7
	10.0	14.0	11.8	---
Intracellular 60K Protein	1.2	---	---	12.4
	2.5	<1	<1	16.3
	5.0	<1	<1	19.1
	10.0	<1	<1	---

* Human G_0B cells isolated from PBLs were activated with 17.5 µg/ml of soluble anti-IgM (Maizel et al., 82) or with 50 µg/ml of dextran sulfate (Okada et al., 83) for 96 hrs. At the end of the activation period, viable cells were isolated by Ficoll-Hypaque gradient centrifugation and used in microcultures for assaying the biological efficacies of both the highly purified low M_r BCGF (Cellular Products Inc., Buffalo) and the homogeneously purified intracellular 60K protein. Both samples were assayed in triplicate at multiple serial dilutions.

These experiments have been reproduced repeatedly using three different donors. The results of a typical experiment are presented here as stimualtion index at each sample concentration tested; the maximum (CPM) [^{3}H]-Tdr incorporated in anti-Igm activated B cells was 42,977 and in dextran sulfate activated cells was 26,852.

\# Stimulation index was calculated as: S.I. = (Experimental mean CPM- background mean CPM)/Background mean CPM. [Table taken from Sahasrabuddhe et al., 86].

TABLE 3

HCL CULTURE SUPERNATANT AUTOSTIMULATORY ACTIVITY

Target Cells	Bkgd	SN (1% BSA)			
		5%	10%	20%	30%
HCL Patient 1	132	990	1263	1725	2752
HCL Patient 2	156	680	1560	2580	ND
BD23	480	981	1867	4332	10175
BDk9	320	684	1424	4427	9564

HCL supernatants (SN) were generated at a final density of 2×10^6 cells per milliliter from E rosette and adherent cell depleted tumor cell populations that were cultured for 24 hours in RPM1 1640 + 1% BSA. Two SN from each patient were obtained and tested. HCL cells are the B cells isolated from the patient's peripheral blood lymphocytes by negative selection procedure described earlier [Ford et al., 81 in Materials and Methods. BD23 and BDk9 are BCGF dependent normal B cell lines. HCL cells (1 x 10^5 >95% viable) or normal B cells (1.5 x 10^4, BD23 or BDk9 cell line) were cultured in microwells for 48 hours with varying dilutions of HCL-derived supernatants. Microwell cultures were labeled with 1 μCi of ^{3}H-Tdr for the final 24 hours of the 72-hour culture period. Results are expressed as mean cpm of triplicate assays. Standard errors were less than 10% [Table taken from Ford et al. 86].

Table 4

Cytoplasmic BCGF activity from HCL Cells

Concentration of growth factor added (% V/V)	[^{3}H]-Tdr incorporation source of cytoplasmic growth repair		
	Normal PBLs	HCL Patient 1	HCL Patient 2
0	198	166	198
0.3	3738+	3740	1620
0.6	7796	4720	2640
1.2	9489	3936	1322
2.5	8962	5396	3633
5.0	ND	6794	4890
10.0	ND	8512	ND
20.0	ND	9484	ND

Target cells for evaluating cytoplasmic BCGF activity consisted of long term B cell lines used at 13000 cells per 200 μl per well.

+ Standard errors were less than 10%
ND -ND indicates not done

DISCUSSION

Homogeneously purified intracellular 60k protein from normal human PBLs supported growth of factor dependent human B cell lines as monitored by cell growth (data not shown) or [^{3}H]-Tdr incorporation in microculture assays, [Table 2]. Yet this protein could not support modulation of s-phase entry of freshly isolated human B cells in either anti-IgM or dextran sulfate costimulatory assays. This protein appeared to be solely B lymphotropic in that it did not exhibit any Interleukin 2 or Interleukin 1 activity (Sahasarabuddhe 85). Results presented here are in contradiction with previously published results (Okada et al 83, Shimizu et al 86, Anonbrus Jr. et al 85). This discrepancy may be attributed to the difference in purification level of the 60K protein, as in the earlier work only partially purified high M_r BCGF secreted by neoplastic lymphocytes was used in the BCGF assays.

The high M_r intracellular 60k protein, a putative precursor to the secreted form of low M_r BCGF (Sahasrabuddhe et al 84), is spatially restricted to the cytoplasmic compartment. Therefore in vivo clonal expansion of antigen activated B lymphocytes is presumably controlled by secreted low M_r extracellular BCGF. Thus, antigen activated normal B cells would not encounter and respond to high M_r intracellular 60k protein. As it is assumed that _in vitro_ anti-IgM activation of human B lymphocytes closely represents a situation analogous to that of _in vivo_ antigen activation, the anti-IgM activated freshly isolated human B lymphocytes would respond to extracellular low M_r (15k) BCGF and not to intracellular high M_r 60k protein. This does not exclude the possibility of anti-IgM activated B cells being responsive to an intermediate form of BCGF. Two putative intermediates of 50k and 30k have been observed by us (Sahasrabuddhe and Martin '87) and by others (Dugas et al 85, Fauci 85). Our results demonstrate that anti-IgM or dextran sulfate activated human B cells respond only to low M_r (15k) BCGF and not to intracellular high M_r 60k protein.

The major source of low M_r BCGF has been the medium conditioned by lectin activated peripheral blood lymphocytes (PBLs), which was also used in the past as a source of human interleukin 2 (IL2). (Mier and Gallo 80). The high M_r BCGF is secreted by T-T hybridomas (Okada et al. 83), neoplastic T lymphocytes (Ambrus Jr. et al. 85), neoplastic B lymphocytes (Ford et al 86, Blazar 83), EBV transformed B lymphocytes (Gordon et al 84, Blazar Sutton, Strome 83) and human T lymphotropic virus transformed T lymphocytes (Shimizu et al 85). Human PBLs when activated simultaneously with PHA and PMA both high and low M_r forms of BCGF (Yoshizaki et al. 83). Studies in the past have shown the specific morphological changes are induced in T lymphocytes including cell membrane damage by PMA (Touraine et al., 77). These changes are likely to cause leakage of unprocessed high mw BCGF from the cytosolic compartment. Recent studies have also shown that human PBLs secrete 50k BCGF when cultured in presence of PHA and sodium azide (10^{-4}M). (Dugas et al 85). Thus it appears that the secretion of 50k or 60 k BCGF molecules is associated with neoplastic and transformed lymphocytes or their release from normal lymphocytes is accompanied with cell lysis or cell death.

Thus it is evident that secretion and biologic efficacy of the 60k protein does not represent a normal mode of BCGF action. Long term B cells (responsive to 60k protein) probably undergo a series of, as yet unknown, biochemical and phenotypic changes before they acquire responsiveness to 60k protein. The observed similarity of responsiveness to 60k protein by neoplastic B cells and normal B cell lines may represent a common mechanism. Two probable changes in B lymphocytes can be speculated at this time. First relates to responsiveness of B cells to 60k protein. Phenotypic changes on the neoplastic B cells may allow the proliferative signal transduction modulated by 60k protein. Secondly, the 60K protein gene which is normally not expressed in B lymphocytes (Sahasrabuddhe et al 84), may be derepressed in neoplastic B cells leading to production of and responsiveness to 60K protein in an autocrine cycle.

ACKNOWLEDGEMENTS

We acknowledge the excellent secretarial assistance of Tammy Trlicek and Linda Jackson. This work was supported in part by National Institutes of Health Grants CA-31479 and GM-35483.

REFERENCES

1. Ambrus JL, Jurgensen CH, Brown EJ, Fauci AS (1985). Purification to homogeneity of a high molecular weight human B cell growth factor: demonstration of specific binding to activated B cells and development of monoclonal antibody to the factor. J Exp Med 162:1319-1335.

2. Blazar BA, Sutton LM, Strome M (1983). Self-stimulating growth factor production by B cell lines derived from Burkitt's lymphomas and other lines transformed in vitro by Epstein-Barr virus. Cancer Res 43:4562-4568.

3. Dugas B, Vazquez A, Gerard JP, Richard Y, Auffredou MT, Delfraissy JF, Fradelizi D, Galanaud P (1985). Functional properties of two human B cell growth factor species separated by lectin affinity column. J Immunol 135:333-338.

4. Ford RJ, Mehta SR, Franzin D, Montagna R, Lachman LB, Maizel AL (1981). Soluble factor activation of human B lymphocytes. Nature 294:261-263.

5. Ford RJ, Kouttab N, Sahasrabuddhe CG, Davis F, Mehta S (1985). Growth factor mediated proliferation in B cell non-Hodgkin's lymphoma. Blood 65:1335-1341.

6. Ford RJ, Kwok D, Quesada J, Sahasrabuddhe CG (1986). Production of B cell growth factor(s) by neoplastic human B cells from hairy cell leukemia patients. Blood 67:573-577.

7. Gillis SM, Ferm M, Ou W, Smith KA (1978). T-cell growth factor: Parameters of production and quantitative microassay for activity. J Immunol 120:2027-2032.

8. Gordon J., Ley SC, Melamed MD, Amen P, Hughes-Jones NC (1984). Soluble factor requirement for the autostimulatory growth of B lymphoblasts immortalized by Epstein-Barr Virus. J Exp Med 159:1554-1559.

9. Jurgensen C, Ambrus Jr JL, Fauci AS (1986). Production of B cell growth factor by normal human B cells. J Immunol 136:4542-4547.

10. Laemmli UK (1970). Determination of protein molecular weight in polyacrylamide gel. Nature 277:680-682

11. Maizel A, Sahasrabuddhe CG, Mehta S, Lachman L, and Ford R (1982). Biochemical separation of human B cell mitogenic factor. Proc Natl Acad Sci (USA). 79:5998-6002.

12. Maizel AL, Mehta SR, Morgan J, Ford RJ, Sahasrabuddhe CG (1983). Proliferation of human B lymphocytes mediated by soluble factor. Fed Proc 42:2753-2756.

13. Maizel AL, Morgan JW, Mehta SR, Kouttab N, Bator J, Sahasrabuddhe CG (1983). Long term growth of human B cells yand their use in a microassay for B cell growth factor. Proc Natl Acad Sci (USA) 80:5047-5051.

14. Mier JW, and Gallo RC (1980). Purification and some characteristics of human T cell growth factor from phytohemagglutinin-stimulated lymphocyte-conditioned media. Proc Natl Acad Sci (USA) 77:6134-6138.

15. Muraguchi A and Fauci AS (1982). Proliferative responses of normal human B lymphocytes: development of assay system for human B cell growth factor (BCGF). J Immunol 129:1104-1108.

16. Oakley BR, Kirsh DR, Morris NR (1980). A simplified ultrasensitive silver stain for detecting proteins in polyacrylamide gel. Annal Biochem 105:361-363.

17. Okada M, Sakaguchi N, Yoshimura N, Hara H, Shimizu K, Yoshida N, Yoshizaki K, Kishimoto T (1983). B cell growth factor (BCGF) and B cell differentiation factor from human T hybridomas: Two distinct kinds of BCGFs and their synergism in B cell proliferation. J Exp Med 157:583-

18. Sahasrabuddhe CG, Morgan J, Sharma S, Mehta S, Martin B, Wright D., Maizel A. (1984). Evidence for an intracellular precursor for human B cell growth factor. Proc Natl Acad Sci (USA) 81:7902-7906.

19. Sahasrabuddhe CG, Dinarello C, Martin B, Maizel A (1985). Intracellular human IL-1: a precursor for the secreted monokine. Lymph Res 4:205-213.

20. Sahasrabuddhe CG, Martin B, Maizel AL (1986). Purification and partial characterization of human intracellular B cell growth factor. Lymph Res 5:127-140.

21. Sahasrabuddhe CG, Martin B (1987). Structural homology between human B cell growth factor and an intracellular 60k protein. Submitted.

22. Touraine JL, Hadden JW, Touraine F, Hadden EM, Estensen R, Good RA (1977). J Exp Med 145:460-465.

23. Yamamura Y, Kishimoto T (1985). Immortalization of BGDF (BCGF II) and BCGF-producing T cells by human T cell leukemia virus (HTLV) and characterization of human BGDF (BCGF II). J Immunol 134:1728-1733.

24. Yoshizaki K, Nagakawa T, Kaieda T, Muraguchi A, Yamamura, Y, and Kishimoto T. (1982). Induction of proliferation and Ig-production in human B leukemic cells by anti-immunoglobulins and T cell factors. J Immunol 130:1241-1246.

THE ROLE OF THE MULTICHAIN IL-2 RECEPTOR COMPLEX IN THE CONTROL OF NORMAL AND MALIGNANT T-CELL PROLIFERATION

Thomas A. Waldmann and Mitsuru Tsudo

The Metabolism Branch, National Cancer Institute, National Institutes of Health, Bethesda, MD 20892

INTRODUCTION

The human body defends itself against foreign invaders such as bacteria and viruses by a defense system that involves antibodies and thymus-derived lymphocytes (T cells). The success of this response requires that human T cells change from a resting to an activated state. Activated T cells are responsible for the regulation of the immune response as well as the elimination of foreign invaders and the rejection of transplanted organs. Failure of the T cells to become activated and function normally may be associated with serious disease and death. The activation of T cells requires two sets of signals from cell surface receptors to the nucleus. The first signal is initiated when appropriately processed and presented foreign antigen interacts with the 90-kd polymorphic heterodimeric T-cell surface receptor for the specific antigen. Following the interaction of the antigen presented in the context of products of the major histocompatibility locus and the macrophage-derived interleukin-1 with the antigen receptor, T cells express the gene encoding the lymphokine IL-2 (Morgan et al., 1976; Smith, 1980). To exert its biological effect, IL-2 must interact with specific high-affinity membrane receptors. Resting T cells do not express IL-2 receptors, but receptors are rapidly expressed on T cells after activation with an antigen or mitogen (Robb et al., 1981; Waldmann, 1986). Thus, the growth factor IL-2 and its receptor are absent in resting T cells, but after activation the genes for both proteins become expressed. Although the interaction of appropriately presented antigen with its specific polymorphic receptor complex confers specificity for a given immune response, the interaction of IL-2 with IL-2 receptors determines its magnitude and duration.

Progress in the analysis of the structure, function, and expression of the human IL-2 receptor was greatly facilitated by our production of the anti-Tac monoclonal antibody that recognizes the human receptor for IL-2 (Uchiyama et al., 1981; Leonard et al., 1982) and blocks the binding of IL-2 to this receptor. Using quantitative receptor binding studies employing radiolabeled anti-Tac and radiolabeled IL-2, it was shown that activated T cells and IL-2-dependent T-cell lines express 5- to 20-fold more binding sites for the Tac antibody than for IL-2 (Depper et al., 1984a,b). Employing high concentrations of IL-2, Robb et al. (1984) resolved these differences by demonstrating two affinity classes of IL-2 receptors. On various cell populations, 5-15% of the IL-2 receptors had a binding affinity for IL-2 in the range of 10^{-11} to 10^{-12} M, whereas the remaining receptors bound IL-2 at a much lower affinity, approximately 10^{-8} or 10^{-9} M. The high-affinity receptors appear to mediate the physiological responses to IL-2 since the magnitude of T cell responses is closely correlated with the occupancy of these receptors. As outlined below, we have utilized the anti-Tac monoclonal antibody and radiolabeled IL-2 in crosslinking studies to: (a) define multiple IL-2 binding peptides that participate in the human receptor for IL-2; (b) molecularly clone cDNAs for the 55-kd peptide of the human IL-2 receptor; (c) analyze disorders of IL-2 receptor expression on leukemic cells; and (d) develop protocols for the therapy of patients with IL-2 receptor-expressing adult T-cell leukemia and autoimmune disorders and for individuals receiving organ allografts.

RESULTS AND DISCUSSION

Chemical Characterization of the IL-2 Receptor

The IL-2 binding receptor peptide identified by the anti-Tac monoclonal on phytohemagglutinin (PHA)-activated normal lymphocytes was shown to be a 55-kd glycoprotein (Leonard et al., 1982, 1983). Leonard and coworkers (1982, 1983) defined the post-translational processing of this 55-kd glycoprotein by employing a combination of pulse-chase and tunicamycin experiments. The IL-2 receptor was shown to be composed of a 33-kd peptide precursor following cleavage of the hydrophobic leader sequence. This precursor was cotranslationally N-glycosylated to 35- and 37-kd forms. After a 1-hour chase, the 55-kd mature form of the receptor appeared, suggesting that O-linked carbohydrate was added to the IL-2 receptor. Furthermore, the IL-2 receptor was shown to be sulfated (Leonard et al., 1984a) and phosphorylated on a serine residue (Shackelford and Trowbridge, 1984).

There were a series of unresolved questions concerning the IL-2 receptor that were difficult to answer when only the 55-kd Tac peptide was considered. These questions included: (a) what is the structural explanation for the great difference in affinity between high- and low-affinity receptors; (b) how, in light of the short cytoplasmic tail of 13 amino acids (see below), are the receptor signals transduced to the nucleus; and (c) how do certain Tac-negative cells (e.g., natural killer cells) make nonproliferative responses to IL-2? To address these questions, we have investigated the possibility that the IL-2 receptor is a complex receptor with multiple peptides in addition to the one identified by anti-Tac. These studies were initiated when we noted that the MLA-144 Gibbon T-cell line binds IL-2 yet does not bind four different antibodies (including anti-Tac and 7G7) that react with different epitopes of the Tac peptide. This cell line manifests 6800 IL-2 receptors per cell with an affinity of 14 nM (Tsudo et al., 1986). On the basis of crosslinking studies using [^{125}I]IL-2, this IL-2-binding receptor peptide was shown to be larger than the Tac peptide with an approximate M_r of 75,000. When similar crosslinking studies were performed on the cell line MT-1 that manifests only low-affinity receptors, only the 55-kd Tac peptide was demonstrated. In contrast, both the 55-kd Tac peptide and the p75 IL-2 binding peptide were expressed on all cell lines (e.g., HUT 102) that manifested both high- as well as low-affinity receptors. Furthermore, fusion of cell membranes from a low-affinity IL-2 binding cell line bearing the Tac peptide alone (MT1) with membranes from a cell line bearing the p75 peptide alone (MLA-144) generated hybrid membranes bearing high-affinity receptors. These studies support a multichain model for the high-affinity IL-2 receptor in which an independently existing Tac or p75 peptide would represent low-affinity receptors, whereas high-affinity receptors would be expressed when both peptides are present and associated in a receptor complex (Tsudo et al., 1986).

T cell functions, including proliferation and differentiation into effectors of suppression and cytotoxicity, are inhibited by anti-Tac and appear to involve the interaction of IL-2 with the multichain high-affinity receptor. In contrast to the action on T cells, anti-Tac does not inhibit the activation of natural killer and lymphokine-activated killer cells by IL-2. In general, such cells are Tac antigen negative but express the p75 IL-2 binding peptide. The interaction of IL-2 with the p75 peptide on natural killer large granular lymphocytes appears to be sufficient to activate such cells to become effective killers, to induce proliferation of such cells, and to induce some such cells to synthesize the Tac peptide.

In addition to the p55 and p75 IL-2 binding peptides, flow cytometric resonance energy transfer measurements support the association of a 93-kd peptide, termed T27, with the 55-kd Tac peptide (Szollosi et al., 1987). The role of this peptide has not been defined.

Molecular Cloning of cDNAs for the Human 55-kd Tac IL-2 Receptor Peptide

Three groups (Cosman et al., 1984; Leonard et al., 1984; Nikaido et al., 1984) have succeeded in cloning cDNAs for the IL-2 receptor protein. The deduced amino acid sequence of the IL-2 receptor indicates that this peptide is composed of 251 amino acids, as well as a 21 amino acid signal peptide. The receptor contains two potential N-linked glycosylation sites and multiple possible O-linked carbohydrate sites. Finally, there is a single hydrophobic membrane region of 19 amino acids and a very short (13 amino acid) cytoplasmic domain. Potential phosphate acceptor sites (serine and threonine, but not tyrosine) are present within the intracytoplasmic domain. However, the cytoplasmic domain of the IL-2 receptor peptide identified by anti-Tac appears to be too small for enzymatic function. Thus, this receptor differs from other known growth factor receptors that have large intracytoplasmic domains with tyrosine kinase activity. However, the p75 peptide associated with the Tac peptide may play a critical role in the transduction of the IL-2 signal to the nucleus.

Disorders of IL-2 Expression in Adult T-Cell Leukemia

A distinct form of mature T-cell leukemia was defined by Takasuki and coworkers (1977) and termed adult T-cell leukemia (ATL). HTLV-I has been shown to be a primary etiologic agent in ATL (Poiesz et al., 1980). All the populations of leukemic cells we have examined from patients with HTLV-I-associated ATL expressed the Tac antigen (Waldmann et al., 1984). The expression of IL-2 receptors on ATL cells differs from that of normal T cells. First, unlike normal T cells, ATL cells do not require prior activation to express IL-2 receptors. Furthermore, using a ^{3}H-labeled anti-Tac receptor assay, HTLV-I-infected leukemic T-cell lines characteristically expressed 5- to 10-fold more receptors per cell (270,000-1,000,000) than did maximally PHA-stimulated T lymphoblasts (30,000-60,000). In addition, whereas normal human T lymphocytes maintained in culture with IL-2 demonstrate a rapid decline in receptor number, adult ATL lines do not show a similar decline. It is conceivable that the constant presence of high numbers of IL-2 receptors on ATL cells may play a role in the pathogenesis of uncontrolled growth of these malignant T cells.

As noted above, T-cell leukemias caused by HTLV-I, as well as all T-cell and B-cell lines infected with HTLV-I, universally express large numbers of IL-2 receptors. An analysis of this virus and its protein products suggests a potential mechanism for this association between HTLV-I and IL-2 receptor expression. The complete sequence of HTLV-I has been determined by Seiki and colleagues (1983). In addition to the presence of typical long terminal repeats (LTRs), gag, pol, and env genes, retroviral gene sequences common to other groups of retroviruses, HTLV-I and -II were shown to contain an additional genomic region between env and the 3' LTR referred to as pX or more recently as tat. Sodroski and colleagues (1984) demonstrated that this pX or tat region encodes a 42-kd protein, now termed the tat protein, that is essential for viral replication. The mRNA for this protein is produced by a double splicing event. These authors demonstrated that the tat protein acts on a receptor region within the LTR of HTLV-I, stimulating transcription. This tat protein may also play a central role in directly or indirectly increasing the transcription of host genes such as the IL-2 and the IL-2 receptor genes involved in T-cell activation and HTLV-I-mediated T-cell leukemogenesis.

The IL-2 Receptor as a Target for Therapy in Patients with ATL and Patients with Autoimmune Disorders and Individuals Receiving Organ Allografts

The observation that ATL cells constitutively express large numbers of IL-2 receptors identified by the anti-Tac monoclonal antibody, whereas normal resting cells and their precursors do not, provides the scientific basis for therapeutic trials using agents to eliminate the IL-2 receptor-expressing cells. Such agents could theoretically eliminate Tac-expressing leukemic cells or activated T cells involved in other disease states while retaining the mature normal T cells and their precursors that express the full repertoire for T-cell immune responses. The agents that have been used or are being prepared include: (a) unmodified anti-Tac monoclonal; (b) toxin (e.g., A chain of ricin toxin, Pseudomonas toxin) conjugates of anti-Tac; and (c) conjugates of alpha-emitting isotopes (e.g., bismuth-212) with anti-Tac.

We have initiated a clinical trial to evaluate the efficacy of intravenously administering anti-Tac monoclonal antibody in the treatment of patients with ATL (Waldmann et al., 1985). None of the seven patients treated suffered any untoward reactions and none produced antibodies to the mouse immunoglobulin or to the idiotype of the anti-Tac monoclonal. Two of the patients had a temporary partial response or complete remission following anti-Tac therapy. In one of these patients, therapy was followed by a 5-month remission, as assessed by routine hematological tests, immunofluorescence analysis of circulating T cells, and molecular genetic analysis of arrangement of the genes encoding the β chain of the T-cell antigen receptor. Following the 5-month remission the patient's disease relapsed, but a new course of anti-Tac infusions was followed by a virtual disappearance of skin lesions and an over 80% reduction in the number of circulating leukemic cells. Two months subsequently, leukemic cells were again demonstrable in the circulation. At this time, although the leukemic cells remained Tac positive and bound anti-Tac in vivo, the leukemia was no longer responsive to infusions of anti-Tac and the patient required chemotherapy. This patient may have had the smouldering form of ATL wherein the leukemic T cells may still require IL-2 for their proliferation. Alternatively, the clinical responses may have been mediated by host cytotoxic cells reacting with the tumor cells bearing the anti-Tac mouse immunoglobulin on their surface by such mechanisms as antibody-dependent cellular cytotoxicity.

These therapeutic studies have been extended in vitro by examining the efficacy of toxins coupled to anti-Tac to selectively inhibit protein synthesis and viability of Tac-positive ATL lines. The addition of anti-Tac antibody coupled to Pseudomonas exotoxin inhibited protein synthesis by Tac-expressing HUT 102-B2 cells, but not that by the Tac-negative acute T-cell line MOLT-4, which does not express the Tac antigen (FitzGerald et al., 1984).

The action of toxin conjugates of monoclonal antibodies depends on their ability to be internalized by the cell and released into the cytoplasm. Anti-Tac bound to IL-2 receptors on leukemic cells is internalized slowly into coated pits and then endosomic vesicles. Furthermore, the toxin conjugate does not pass easily from the endosome to the cytosol, as required for its action on elongation factor 2. To circumvent these limitations, an alternative cytotoxic reagent was developed that could be conjugated to anti-Tac and that was effective when bound to the surface of leukemic cells. It was shown that bismuth-212 (^{212}Bi), an alpha-emitting radionuclide conjugated to anti-Tac by use of a bifunctional chelate, was well suited for this role (Kozak et al., 1986). Activity levels of 0.5 μCi or the equivalent of 12 rad/ml of alpha radiation targeted by ^{212}Bi-anti-Tac eliminated greater than 98% of the proliferative capacity of the HUT 102-B2 cells, with only a modest effect on IL-2 receptor-negative lines. This specific cytotoxicity was blocked by excess unlabeled anti-Tac, but not by human IgG. Thus, ^{212}Bi-anti-Tac is a potentially effective and specific immunocytotoxic agent for the elimination of IL-2 receptor-positive cells.

In addition to its use in the therapy of patients with ATL, antibodies to the IL-2 receptors are being evaluated as potential therapeutic agents to eliminate activated IL-2 receptor-expressing T cells in other clinical states, including certain autoimmune disorders and in protocols involving organ allografts. The rationale for the use of anti-Tac in patients with the disease aplastic anemia is derived from the work of Zoumbos and coworkers (1985) who have demonstrated that select patients with aplastic anemia have increased numbers of circulating Tac-positive cells. In this group of patients, the Tac-positive but not Tac-negative T cells were shown to inhibit hematopoiesis when cocultured with normal bone marrow cells. Furthermore, we have demonstrated that anti-Tac inhibits the generation of activated suppressor T cells (Oh-ishi and Waldmann, unpublished observations). Studies have been initiated to define the value of anti-Tac in the therapy of patients with aplastic anemia. The rationale for the use of an antibody to IL-2 receptors in recipients of renal and cardiac allografts is that anti-Tac inhibits the proliferation of T cells to foreign histocompatibility antigens expressed on the donor organs and prevents the generation of cytotoxic T cells in allogeneic cell cocultures. Furthermore, in studies by Strom and coworkers (1985), the survival of renal and cardiac allografts was prolonged in rodent recipients treated with an anti-IL-2 receptor monoclonal antibody. Thus, the development of monoclonal antibodies directed toward the IL-2 receptor expressed on ATL cells, on autoreactive T cells of certain patients with autoimmune disorders, and on host T cells responding to foreign histocompatibility antigens on organ allografts may permit the development of rational new therapeutic approaches in these clinical conditions.

SUMMARY

Antigen-induced activation of resting T cells induces the synthesis of interleukin-2 (IL-2), as well as the expression of specific cell surface receptors for this lymphokine. There are at least two forms of the cellular receptors for IL-2, one with a very high affinity and the other with a lower affinity. We have identified two IL-2 binding peptides, a 55-kd peptide reactive with the anti-Tac monoclonal antibody and a 75-kd non-Tac IL-2 binding peptide. Cell lines bearing either the p55, Tac, or the p75 peptide alone manifested low-affinity IL-2 binding, whereas cell lines bearing both peptides manifested both high- and low-affinity receptors. Fusion of cell membranes from low-affinity IL-2 binding cells bearing the Tac peptide alone with membranes from a cell line bearing the p75 peptide alone generated hybrid membranes bearing high-affinity receptors. We propose a multichain model for the high-affinity IL-2 receptor in which both the p55 Tac and the p75 IL-2 binding peptides are associated in a receptor complex. The p75 peptide is the receptor for IL-2 on large granular lymphocytes and is sufficient for the IL-2 activation of these cells. In contrast to resting T cells, human T-cell lymphotropic virus I-associated adult T-cell leukemia cells constitutively express large numbers of IL-2 receptors. Because IL-2 receptors are present on the malignant T cells but not on normal resting cells, clinical trials have been initiated in which patients with adult T-cell leukemia are being treated with either unmodified or toxin-conjugated forms of anti-Tac monoclonal antibody directed toward this growth factor receptor.

REFERENCES

Cosman D, Cerreti DP, Larsen A, Park L, March C, Dower S, Gillis S, Vidal D (1984). Cloning, sequence and expression of human interleukin-2 receptor. Nature 321:768-771.

Depper JM, Leonard WJ, Krönke M, Waldmann TA, Greene WC (1984a). Augmentation of T-cell growth factor expression in HTLV-I-infected human leukemic T cells. J Immunol 133:1691-1695.

Depper JM, Leonard WJ, Krönke M, Noguchi P, Cunningham R, Waldmann TA, Greene WC (1984b). Regulation of interleukin-2 receptor expression: Effects of phorbol diester, phospholipase C, and reexposure to lectin and antigen. J Immunol 133:3054-3061.

FitzGerald D, Waldmann TA, Willingham MC, Pastan I (1984). Pseudomonas exotoxin-anti-Tac: Cell specific immunotoxin, active against cells expressing the T-cell growth factor receptor. J Clin Invest 74:966-971.

Kozak RW, Atcher RW, Gansow OA, Friedman AM, Waldmann TA (1986). Bismuth-212 labeled anti-Tac monoclonal antibody: Alpha-particle emitting radionuclides as novel modalities for radioimmunotherapy. Proc Natl Acad Sci USA 83:474-478.

Leonard WJ, Depper JM, Uchiyama T, Smith KA, Waldmann TA, Greene WC (1982). A monoclonal antibody that appears to recognize the receptor for human T cell growth factor: Partial characterization of the receptor. Nature 300:267-269.

Leonard WJ, Depper JM, Robb RJ, Waldmann TA, Greene, WC (1983). Characterization of the human receptor for T cell growth factor. Proc Natl Acad Sci USA 80:6957-6961.

Leonard WJ, Depper JM, Waldmann TA, Greene WC (1984a). A monoclonal antibody to the human receptor for T cell growth factor. In Greaves M (ed): "Receptors and Recognition," Vol. 17, London: Chapman and Hall, pp 45-66.

Leonard WJ, Depper JM, Crabtree GR, Rudikoff S, Pumphrey J, Robb RJ, Krönke M, Svetlik PB, Peffer NJ, Waldmann TA, Greene WC (1984b). Molecular cloning and expression of

cDNAs for the human interleukin-2 receptor. Nature 311:626-631.
Morgan DA, Ruscetti FW, Gallo, RC (1976). Selective in vitro growth of T lymphocytes from normal human bone marrows. Science 193:1007-1008.
Nikaido T, Shimizu N, Ishida N, Sabe H, Teshigawara K, Maeda M, Uchiyama T, Yodor S, Honjo T (1984). Molecular cloning of cDNA encoding human interleukin-2 receptor. Nature 311:631-635.
Poiesz BJ, Ruscetti FW, Gazdar AF, Bunn PA, Minna JD, Gallo RC (1980). Detection and isolation of type C retrovirus particles from fresh and cultured lymphocytes of a patient with cutaneous T-cell lymphoma. Proc Natl Acad Sci USA 77:7415-7419.
Robb RJ, Munck A, Smith KA (1981). T-cell growth factors: Quantification, specificity, and biological relevance. J Exp Med 154:1455-1474.
Robb RJ, Greene WC, Rusk CM (1984). Low and high affinity cellular receptors for interleukin 2: Implications for the level of Tac antigen. J Exp Med 160:1126-1146.
Seiki M, Hattori S, Hirayama Y, Yoshida M (1983). Human adult T-cell leukemia virus: Complete nucleotide sequence of the provirus genome integrated in leukemic cell DNA. Proc Natl Acad Sci USA 80:3618-3622.
Shackelford DA, Trowbridge IS (1984). Induction of expression and phosphorylation of the human interleukin 2 receptor by a phorbol diester. J Biol Chem 259:11706-11712.
Smith KA (1980). T-cell growth factor. Immunol Rev 51:337-357.
Sodroski JG, Rosen CA, Haseltine WA (1984). Transacting transcriptional activation of the long terminal repeat of human T lymphotrophic viruses in infected cells. Science 225:381-385.
Strom TB, Banet LV, Gauiton GN, Kelley VF, Thier AY, Diamanstein T, Tilney NL, Kirkman RL (1985). Prolongation of cardiac allograft survival in rodent recipients treated with an anti-interleukin-2 receptor monoclonal antibody. Cancer Res 33:561A.
Szollosi J, Damjanovich S, Goldman CK, Fulwyler M, Aszalos AA, Goldstein G, Rao P, Talle MA, Waldmann TA (1987). Flow cytometric resonance energy transfer measurements support the association of a 93-kDa peptide termed T27 with the 55-kDa Tac peptide. In press.
Takasuki K, Uchiyama T, Sagawa K, Yodoi J (1977). Adult T cell leukemia in Japan. In Seno S, Takaku F, Irino S (eds): "Topics in Hematology," Amsterdam: Excerpta Medica, pp 73-77.

Tsudo M, Kozak RW, Goldman CK, Waldmann TA (1986). Demonstration of a new (non-Tac) peptide that binds interleukin-2: A potential participant in a multichain interleukin-2 receptor complex. Proc Natl Acad Sci USA 83:9694-9699.

Uchiyama T, Broder S, Waldmann TA (1981). A monoclonal antibody (anti-Tac) reactive with activated and functionally mature human T cells. J Immunol 126:1393-1397.

Waldmann TA (1986). The structure, function, and expression of interleukin-2 receptors on normal and malignant T cells. Science 232:727-732.

Waldmann TA, Greene WC, Sarin PS, Saxinger C, Blayney W, Blattner WA, Goldman CK, Bongiovanni K, Sharrow S, Depper JM, Leonard W, Uchiyama T, Gallo RC (1984). Functional and phenotypic comparison of human T cell leukemia/lymphoma virus positive adult T cell leukemia with human T cell leukemia/lymphoma virus negative Sezary leukemia. J Clin Invest 73:1711-1718.

Waldmann TA, Longo DL, Leonard WJ, Depper JM, Thompson CB, Krönke M, Goldman CK, Sharrow S, Bongiovanni K, Greene WC (1985). Interleukin-2 receptor (Tac antigen) expression in HTLV-I associated adult T-cell leukemia. Cancer Res 45:4559s-4562s.

Zoumbos N, Gascon P, Trost S, Djeu J, Young N (1985). Circulating activated suppressor T lymphocytes in aplastic anemia. N Engl J Med 312:257-265.

Index

Actinomycin D, 168, 169, 242, 247, 248
Adrenal androgens, 52
    after castration, 17–18, 19
    and prostatic cancer, 11–12, 13–14
    tumor cell sensitivity, 20
Adrenal cortical cell lines, 202
Adrenalectomy, 18, 19, 20, 53
A-431, 210–211
Aldosterone, 250
Amino acid sequence, estrogen receptor, 182
Aminoglutethimide, 76, 78, 224–226
    as aromatase inhibitor, 70
    in prostatic cancer, 12, 18, 19, 20
Anandron, 29
Androgen receptors, 79
    antiandrogens and, 52–53
    aromatase inhibitors and, 74, 75
    assay, 145–158
Androgens, in prostatic cancer, 11–17
    after castration, intraprostatic levels, 16–18
    tumor sensitivity to, 19–22
    see also Prostate cancer
$3\alpha,17\beta$-Androstanediol, 137
5α-Androstane-$3\alpha,17\beta$-diol, 17
5α-Androstane-$3\beta,17\beta$-diol, 137
Androst-5-ene-$3\beta,17\beta$-diol, 50, 54
Androstenedione, 50, 54, 66
    aromatase reactions, 67
    and benign prostatic hypertrophy, 137
Antiandrogen
    LHRH agonists vs., 224–226
    see also Prostate cancer
Antiestrogens
    in breast cancer, 105–116
    and estrogen receptor, chromatin changes, 85–96
Antiglucocorticoid, 170
Antiprogestins, 168, 170, 171–175
Anti-Tac antibody, 284–291
Aplastic anemia, 289

Ara C, 169
Aromatase
    characteristics
        in vitro studies, 70–76
        in vivo studies, 76–78
    inhibitors of
        in vitro studies, 70–76
        in vivo studies, 76–78
        in benign prostatic hypertrophy, 137
Atherosclerosis, 114

B cell growth factor, 269–279
Benign prostatic hypertrophy, 125–139
Breast cancer
    aromatase and inhibitors, 65–79
    estrogen effects, cellular, 181–189
        macromolecules, 183–184
        receptor, 182–183
    estrogen therapy, 7–8
    LHRH analogs in, 231–234
    tamoxifen and, 105–116
Breast cancer cell line T47D, 161–176
B16/C3 cells, 241–253
Buserelin, 224–226, 233
t-Butylphenoxyethyl diethylamine, 87

Castration, medical. *See* Antiandrogens
Castration, surgical. *See* Orchiectomy
Cell culture
    epidermal growth factor binding, 210–211
    hormonally defined media, 199–206
    melanoma cells (B16/C3), 241–252
Cell proliferation, steroids and, 3–8
Cellular endocrinology, 199–206
Chemotherapy
    breast cancer, 112
    prostate cancer, 226–230
Chick oviduct, 89
Chlormadinone acetate, 12, 23, 24, 25
Cholecystokinin, 234
Cholera toxin, 242

Chondrosarcomas, 230–231
Chromatin, estrogen and antiestrogen effects, 85–96
Chromosomes, *Drosophila*, 3–4
Cloudman S91 cell line, 242
Conalbumin, 89
Contraceptive pills, 8
Coronary artery disease, 114
Cortisol, 50, 153
Cyclic AMP, 241, 242
Cycloheximide, 168, 242, 247
Cyclophosphamide, 226–228
Cyproterone acetate, 22, 23, 24
    vs. flutamide combination therapy, 35–39
    in stage C patients, 43, 44
    treatment failure, 49, 50
Cytosine-β-D-arabinofuranoside, 169
Cytoxan, 226–228

Dehydroepiandrosterone, 50, 54
Dehydroepiandrosterone sulfate, 50
Deoxycorticosterone, 250
DES, in prostate cancer, 12, 14
    and androgen levels, 17
    vs. antiandrogens, 25
    and estrogen receptor, prostate, 127, 128, 130, 131, 132
    vs. flutamide combination therapy, 33–39
    in stage C patients, 43, 44, 45
    treatment failure, 49, 50
Dexamethasone, 250
Differentiation
    leukemia cells, 257–266
    steroids and, 3–8
    see also Growth factors
Dihydrofolate reductase, 183
5α-Dihydrotestosterone
    in prostate cancer, 148
        androgen receptor assay, 153
        antiandrogen and, 11
        after castration, 16
        flutamide and, 52
    and tyrosinase, 250
Diisopropylfluorophosphate, 137
Dimethylbenzanthracene, 107
Dimethyl-19-nortestosterone, 146, 148–158
Dithiothreitol, 126, 129, 130, 135
Divalent cations, 139
Drosophila melanogaster
    development, 3–6
    EGF receptor-related protein, 210
Dunning tumor, 219–222, 224–226, 231

Ecdysone, 3–6
EDTA, and prostatic estrogen receptor, 126, 131, 135–136, 137
Endometrial cancers, 8
Endothelial cell growth factor, 187
Enzymes
    aromatase. *See* Aromatase
    estrogens and, 183
    tyrosinase. *See* Tyrosinase
Epidermal growth factor
    assay of, 209
    in breast cancer cells, 185
    receptor, 210–213
    sequence comparisons, 208–209
erb-A gene family, 183
    sequence homologies, 182
    steroid receptors and, 7, 181
v-*erb*-B, epidermal growth factor and, 185, 212
Erythropoietin, 257–266
Estracyt, 44, 45, 49, 50
Estradiol, 168
    aromatase and. *See* Aromatase
    and benign prostatic hypertrophy, 137
    prostatic androgen receptor assay, 153
    tamoxifen and, 105
    and tyrosinase, 248–250
Estradiol-17b, 168
Estriol, 67
Estrogen, 79
    antiestrogen and, 114
    aromatase and. *See* Aromatase
    in breast cancer
        cellular effects, 181–189
        hormonal therapy approaches, 7–8
    in prostatic cancer, 13, 18
        acceptability of, 15–16
        androgen effects, 17
    and androgen receptors, 54
    flutamide combination therapy vs., 33–34
    LHRH analogs and, 53, 232
    insensitivity, 19–21
    vs. other therapies, 39, 40
    side effects, 56–57

and tyrosinase, 248–250
Estrogen receptor
   aromatase inhibitors and, 74, 75
   in breast cancer, 7–8, 105–106, 182, 183
   estrogen and antiestrogen effects, 85–96
   prostate, assay of, 125–139
Estrogen receptor messenger RNA, 182
Estrone, 66, 67

Fibroblast growth factor, 187
5'-p-Fluorosulfonylbenzoyl adenosine, 211
Flutamide, in prostate cancer, 29
   adrenal androgens, 11–12, 13
   androgen sensitivity, range of, 19–22
   antiandrogen properties, 22–24
   castration
      androgen-sensitive tumor cells after, 18–19
      dihydroxytestosterone after, 16–18
   LHRH agonists vs., 224–226
   relapsing patients, 13–14, 45–47
   results of treatment, 48–57
   stage C patients, 41–44
   stage D2 patients, previously untreated, 25–41

Gastrin, 234
GH3, 202, 203, 204
Glial cell lines, 202
Glucocorticoid receptor, 182
Glucocorticoids
   chondrosarcoma and, 230, 231
   tamoxifen and, 89
Growth factors
   in breast cancer
      autocrine-stimulated growth, 184–189
      epidermal growth factor and transforming growth factor α, 185–186
      insulin-like growth factor, 185
      in vivo effects of growth factors, 187–188
      other growth factors, 186–187
      platelet-derived growth factor, 184–185
      ras oncogenes, 188–189
      transforming growth factor beta, 187
   epidermal growth factor, 207–213
   erythropoietin, 257–266

IL-2 receptor, 283–291
Growth hormone
   cell culture systems, 202, 203, 204
   chondrosarcoma and, 230, 231
   LHRH analog and, 231
   progestins and, 162
   in prostatic cancer, 222

Hematopoiesis
   erythropoietin, 257–266
   Tac-negative T cells and, 289–291
HTLV-1, 287
Human epidermoid carcinoma cell, EGF binding, 210–211
Human pancreatic secretory trypsin inhibitor, 208
Hydronephrosis, 26, 38, 43
4-Hydroxyandrostenedione, 70–79, 137
16-Hydroxyandrostenedione, 67
19-Hydroxyandrostenedione, 66
17β-Hydroxysteroid dehydrogenase, 161
16-Hydroxytestosterone, 67
Hypophysectomy, 18, 19, 20, 53
Hypothalamic hormones, 218

Imidazole
   steroid hormones and, 248, 249, 250
   and tyrosinase, 243, 244–245, 246
Immune system
   B cell growth factor, 269–279
   IL-2 receptor, 283–291
Insulin-like growth factors, 185
Interleukin-2 receptor, 283–291

Lactate dehydrogenase, progestin and, 161–176
Leukemias
   B cell growth factor, 269–279
   erythropoietin, 257–266
   IL-2 receptor, 283–291
Leuprolide, 12, 32–35, 233
Leydig cell lines, 202
LHRH analogs
   in breast cancer, 231–234
   in chondrosarcomas and osteosarcomas, 230–231
   in pancreatic cancer, 234–235
   in prostatic cancer, 218–222

with antiandrogen, 224–226
with cytoxan, 226–228
with chemotherapy, 226–230
Dunning tumor, 219–222
human studies, 223–224
with Novantrone, 226–230
with somatostatin analogs, 222–223
in prostatic cancer, combination therapy with flutamide, 11–57
adrenal androgens, 11–12, 13
androgen sensitivity, range of, 19–22
antiandrogen properties, 22–24
castration effects, 16–19
relapsing patients, 13–14, 45–47
results of treatment, 48–57
stage C patients, 41–44
stage D2 patients, previously untreated, 25–41
Lymphocytes
B cell growth factor, 269–279
IL-2 receptor, 283–291

Mammary carcinoma models. *See* Breast cancer
Mammary-derived growth factor, 187
Maturation, leukemia cells, 257–266
MCF-7 cells, 107, 108, 162
growth factors, 186, 187
in vivo tumor formation, 187–188
phenol red and, 171
progestin effects, 162, 164, 165, 168
MDA-MB-231, 186
Medroxyprogesterone acetate, 23, 24, 25
as antiandrogen, 12
vs. flutamide combination therapy, 35–39
in stage C patients, 43, 44
Megestrol acetate, 12, 22, 23, 24, 25
Melanocyte-stimulating hormone (MSH), 241
Melanoma cells, tyrosinase in, 241–252
steroids and, 248–250
thyroid hormone, 243–248
Mibolerone, 146, 148–158
Mitoxantrone, 228–230
Molybdate, 126, 130, 131, 132, 135–136, 137

N-Nitroso-methylurea, 108
Nolvadex. *See* Tamoxifen

Novantrone, 228–230

Oral contraceptives, 8
Orchiectomy
acceptability of, 15–16
adrenal androgens after, 11–12, 13
androgen-hypersensitive tumors after, 18–19
androgen levels after, 16–18
antiandrogens with, 22
effectiveness of, 19–21
flutamide vs., 33–34
flutamide with, 53
vs. LHRH agonists, 223
LHRH agonists as substitutes, 53
vs. other therapies, 39, 40
relapse after, 13–14
stage C patients, 44, 45
Ornithine decarboxylase, 24, 25
Osteoporosis, 114
Osteosarcomas, 230–231
Ovarian ablation, 113
Ovarian carcinoma, 8, 79
Ovary
aromatase in, 65
LHRH analogs and, 232
tamoxifen and, 114

Pancreatic cancer, 234–235
Parathyroid cells, 204–205
D-Phe-Cys-Tyr-D-Trp-Lys-Val-Cys-Trp-NH$_2$, 222
Phenol red, 171–172, 173, 174
Phenylmethylsulfonyl fluoride, 126, 132–133, 134
Phosphatidylinositol, 212–213
Pituitary
cell lines, 202, 203, 204
LHRH analogs and, 232
Pituitary growth factor, 187
Platelet-derived growth factor, 184–185
Progesterone
LHRH analogs and, 231, 232
prostatic androgen receptor assay, 153
and tyrosinase, 250
Progesterone receptor
EDTA and, 136
sequence homologies, 182
Progestins

in breast cancer, in vitro effects, 161–176
oral contraceptives, 8
tamoxifen and, 89
Prolactin
  LHRH analog and, 231
  progestins and, 162
  in prostatic cancer, 222
10-Propargylestr-4-ene-3,17-dione (PED), 70–79
Prostaglandin E1, 242
Prostate, 8
  androgen receptor assay, 145–150
  estrogen receptors, 125–139
Prostate cancer
  LHRH analogs
    with antiandrogen, 224–226
    with cytoxan, 226–228
    with chemotherapy, 226–230
    Dunning tumor, 219–222
    human studies, 223–224
    with Novantrone, 226–230
    with somatostatin analogs, 222–223
  LHRH analogs with flutamide, 11–57
    adrenal androgens, 11–12, 13
    androgen sensitivity, range of, 19–22
    antiandrogen properties, 22–24
    castration effects, 16–19
    relapsing patients, 13–14, 45–47
    results of treatment, 48–57
    stage C patients, 41–44
    stage D2 patients, previously untreated, 25–41
Prostatic acid phosphatase, 30
Protein kinase C, 213
Protein kinases, 211–212, 213

ras oncogenes, 180
RC-15, 233
RC-160, 222, 233
R5020, 165–166, 168–175
R1881, 153
Rous sarcoma virus-transforming protein, 182
RU486, 168, 170, 171–175
RU23901, 224–226
RU23908, 29

Secretin, 234
Serine protease inhibitors, 137

Shionogi tumor, 20–21, 55
c-*sis* gene, 184
Somatomedin, 230, 231
Somatostatin analogs
  in breast cancer, 233
  in chondrosarcomas and osteosarcomas, 230–231
  in pancreatic cancer, 234–235
  in prostatic cancer, 222–224
Spironolactone, 24, 25
Steroid hormones
  aminoglutethimide and, 76
  and tyrosinase, 248–250
  see also specific hormones
Swarm chondrosarcoma, 230, 231
SW13, 186

Tac peptide, 284–291
Tamoxifen, 176
  in breast cancer, 105–116
    adjuvant therapy clinical evaluation, 108–113
    pharmacological basis, 107
    prevention, 113–115
  and estrogen receptor, chromatin changes, 85–96
T cells, IL-2 receptor, 283–291
Teratoma, 202
Testolactone, 70
Testosterone
  aromatase inhibitors and, 77
  after castration, 11–12, 13–14
  and estrogen receptors, prostatic, 138
  LHRH analogs and, 223
T47D, 161–176
Thyroid cell lines, 204
Thyroid hormone, and tyrosinase, 243–248
Thyroid hormone receptor, c-*erb*-A as, 183
Transforming growth factor alpha, 185, 208
Transforming growth factor beta, 187
D-Trp-6-LHRH
  with somatostatin analogs, 222–223
  see also LHRH analogs
Tryptex. *See* LHRH analogs
Tyrosinase, 241–252
Tyrosine kinase, 211–212

Urinary obstruction, 13, 26
Uterine estrogen receptor, 105
Uterotrophic activity, tamoxifen, 89

Vaccinia growth factor, 208

Zinc, in prostate, 135, 139

THE LIBRARY
UNIVERSITY OF CALIFORNIA
San Francisco